Working with the Past: Towards an Archaeology of Recycling

Edited by

Dragoș Gheorghiu and Phil Mason

ARCHAEOPRESS ARCHAEOLOGY

Archaeopress Publishing Ltd
Gordon House
276 Banbury Road
Oxford OX2 7ED

www.archaeopress.com

ISBN 978 1 78491 629 9
ISBN 978 1 78491 630 5 (e-Pdf)

© Archaeopress and the individual authors 2017

Cover: First cover image: Albesti an early 20th century house built with stones from the local Graeco-Roman settlement (Photo by D. Gheorghiu)

All rights reserved. No part of this book may be reproduced, in any form or by any means, electronic, mechanical, photocopying or otherwise, without the prior written permission of the copyright owners.

Printed in England by Oxuniprint, Oxford
This book is available direct from Archaeopress or from our website www.archaeopress.com

Contents

List of Figures and Plates iii

Contributors vii

The Never Ending Journey: Cycling and Recycling Seen through a Critical Assessment of the Taphonomic Process 1
Roberta Robin Dods

Sustainability, Health, and Society: Prehistoric Artefacts as Sustainable Materials 19
Lolita Nikolova

Recycling Power and Place: The Many Lives of Traprain Law, SE Scotland 27
Ian Armit, Andrew Dunwell, Fraser Hunter

Tells as Recycled Places. Experimenting the Chalcolithic Ritual Technologies of Construction and Deconstruction 37
Dragoş Gheorghiu

Copper and Bronzes: The Birth of Complete Recycling in The Bronze Age 49
Davide Delfino

Rock Art Recycled? On the Use of Bronze Age Rock Art Sites during the Iron Age in Southern Scandinavia 63
Per Nilsson

Recycled Memories: The Past and Present in Early Iron Age Landscapes of Southern Germany 77
Matthew L. Murray

Ancestral Places: The Creation and Recycling of Monumental Landscapes in South-Eastern Slovenia in The 1st Millennium BC and the 1st Millennium AD 87
Phil Mason

Recycling Pots, Places and Practices: The Roman Cemetery at Podlipoglav 97
Bernarda Županek and Irena Sivec

Secondary Use of Storage Vessels and Household Pottery During the Late Middle Ages: Pottery in Vaults as a Case Study 105
Marta Caroscio

The Reuse of Materials during the Medieval and Post-Medieval Periods: A Case Study of Recycling Building Materials in Rothwell, near Leeds, England 111
George Nash

List of Figures and Plates

The Never Ending Journey
Roberta Robin Dods

Plate 1: Item of war turned into a child's toy. An example of use and reuse from a Bedouin camp in Jordan. Image R.R. Dods 1990......1

Fig. 1: Karl Popper's Three Worlds of Knowledge. (http://www.knowledgejump.com/knowledge/popper.html) 3

Fig. 2: Flow model for archaeological materials adapted and expanded from Schiffer (1972: 158-159), Lange and Rydberg (1972), and Clarke (1968: 36). Note the material remains of subject culture (1) in systemic contexts of GROUP 2 and GROUP 3 (the archaeologists). Recycle is recycle regardless!........6

Fig. 3: Model of the linkages/relationships of taphonomies I through IV. The main feature is diminishment of information. Three dimensions of space (in the diagram represented by the cube) and the dimension of time (represented by the movement into the visual foreground of the model on the page) are depicted. 6

Chart 1: Contexts and their meaning for Figs. 2 and 3........7

Plate 2: Recycling discarded items witin a culture (potential Group 1 returned to Group 1). (Hudson 2009) 10

Plate 3: Dumpster diving. Recycling of food from discard area. Julia Golomb (left) and Alison Abreu-Garcia, both of Somerville, mine a dumpster of a metro-area grocery store. (Baker 2009 with Globe photo by Gretchen Ertl) 11

Plate 4: Mahalapye, Botswana traditional house with walled 'front yard'. (http://en.wikipedia.org/wiki/File:Mahalapye_traditional_house_ cropped.jpg accessed 22.01.2017) 11

Sustainability, Health, and Society
Lolita Nikolova

Scheme 1: Societal components related to sustainability: production (1), repairing (2), secondary use (3), recycling (4) and wasting (5). 19

Scheme 2: Prehistoric fragmented pottery as a sustainable material: non-building (A), foundation of paths (B), and building material (C)........20

Scheme 3: Explanatory models of obtaining of earlier prehistoric fragmented pottery for use in later levels: by digging pits in the village (A), digging ditches in or around the village (B), obtaining soil from the periphery of the village (C), rituals (D) or other (E). 20

Scheme 4: Modelling of possible reasons for finding fragmented pottery in prehistoric levels........22

Recycling Power and Place
Ian Armit, Andrew Dunwell, Fraser Hunter

Fig. 1: Location map (drawn by Rachael Kershaw)........28

Fig. 2: Traprain Law, East Lothian (photo: Ian Armit)........28

Fig. 3: Simplified plan of Traprain Law showing the main focal areas of excavation (drawn by Libby Mulqueeny) 29

Fig. 4: AMS dates from Traprain Law, funded by Historic Scotland, calibrated using Oxcal 4 (Bronk Ramsey 2009, Reimer et al. 2004)........32

Fig. 5: The axe hoard found in the burnt out area of Traprain Law in 2004 (National Museums of Scotland)........33

Fig. 7: Stone 'plaque' made from fragment of the linear rock art (National Museums of Scotland) 33

Fig. 6: The axe hoard found in the burnt out area of Traprain Law in 2004 (National Museums of Scotland)........33

Fig. 8: North Berwick Law framed in the out-turned entrance-way to the innermost enclosure at Traprain Law (photo: Ian Armit)........34

Tells as Recycled Places
Dragoş Gheorghiu

Fig. 1: The experimental reconstruction by the author of the first level of dwelling in a Chalcolithic settlement (Vădastra 2003). 38

Fig. 2: Uzunu tell used as a clay quarry by villagers (2007)........39

Fig. 3: Fishermen houses at the base of Hârşova tell (2005)........39

Fig. 4: The building of a wattle and daub house (2003). 40

Fig. 5: Fragment of a burned wall with visible vegetal straws. (Sultana tell). 40

Fig. 6: A foundation trench (Vădastra experiments 2003). 41

Fig. 7: The ceramic wall of a burned house (Vădastra 2006). 42

Fig. 8: Unburned part of a structural post (Vădastra 2010). 42

Fig. 9: The burned down house three years after the collapse (Vădastra 2009). 43

Fig. 10: An unburned wattle and daub house left eight years to weathering with a recyclable wooden structure (Vădastra 2011). 43

Fig. 11: A mass of large burned architectural fragments (Uzunu tell, 2007) 44

Fig. 12: Several layers of burned dwellings separated by levelling layers (Hârşova tell, 2005) 45

Copper and Bronzes
Davide Delfino

Fig. 1: Investigated areas; Alpine region (A) and Atlantic Iberian Peninsula (B) 52

Fig. 2: Alpine region: the copper resources (black circles) and cited sites (1. Baragalla; 2. Casse Rousse; 3. Pinerolo; 4. Bric della Sorte; 5. St. Pierre d' Albigny; 6. Meytet; 7.Lugana Vecchia; 8. Castellarano; 9. Frattesina di Fratta Polesine; 10. S. Francesco di Bologna) (Elaborate from Cierny 1997: 77) 52

Fig. 3: Bric della Sorte hoard (Savona-Liguria) (Image of Soprintendenza per i Beni Archeologici della Liguria). 53

Fig. 4: Atlantic Iberian Peninsula: the copper (circles) and tin (triangles) resources and citied sites (1. Quinta do Ervedal; 2. Casal dos Fieis de Deus; 3. Vila Cobva de Perrinho; 4. Ria Huelva; 5. Castro de Nossa Senhora de Guia; 6. Castelejo; 7/8. Alegrios/ Moreirinha; 9. Monte do Frade) (Elaborate from Cardoso, Vilaça 2008: 44). 55

Fig. 5: Porto do Concelho hoard (Mação- Pinhal Interior). 55

Fig. 6: Lugana Vecchia scrap in settlement (Brescia- Lombardia) (Elaborate from De Marinis 2006a: 1298,1300). 56

Fig. 7: Frattesina di Fratta Polesine scrap in metallurgical workshop No. 4 (Padova-Veneto). 57

Rock Art Recycled?
Per Nilsson

Fig. 1: Map showing the location of the Himmelstalund region. 63

Fig. 2: Map showing the rock art site at Himmelstalund with the location of the hearths and the settlement remains. 65

Fig. 3: A house from the Early Iron Age was found between the rock art site and the nearby river Motala Ström. Photo Per Nilsson. 66

Fig. 4: Two hearths were found beneath one of the panels at Himmelstalund. Photo Per Nilsson. 66

Fig. 5: Rock art covered with fire cracked stones at Leonardsberg. From Nordén 1925. 67

Fig. 6: The cemetery at Fiskeby. Map by Per Lundström, copy from the Royal Swedish Academy of Letters, History and Antiquities. 68

Fig. 7: At Lille Strandbygård on the Danish island of Bornholm, two houses were found beside the rock art. From Sørensen 2006: 72. 69

Fig. 8: The runic inscription from Himmelstalund. Photo Per Nilsson. 70

Recycled Memories
Matthew L. Murray

Fig. 1: Location of Tumulus 17 (excavated in 1999-2000 as part of the Landscape of Ancestors project) in the Speckhau (Hohmichele) mound group near the Heuneburg. The map shows mounds and mound groups that are traditionally considered part of the Heuneburg mortuary landscape (adapted from Kurz, S. 2007, Figure 4). 79

Fig. 2: Plan view of Tumulus 17 showing the remains of the primary cremation grave (Gr. 5) in the central enclosure and the location and orientation of secondary graves. 79

Fig. 3: Photograph of the south and east profiles of the northwest quarter of Tumulus 17 in 1999 showing mound stratigraphy. The distinction between the original (inner) mound and the later (outer) mound layers can be clearly seen (author's photograph). 80

Fig. 4: Idealized profile of Tumulus 17 showing original (inner) and later (outer) mound fill and the location of refit pottery fragments (pottery refit data from Schneider 2003: Figure 56). 80

Fig. 5: Plan of the structured landscape of the Heuneburg, including the hillfort and additional earthworks, as well as the "chamber" gate and its orientation toward the burial mounds at Gießübel-Talhau. Ditches are shown in grey and wall remnants are shown in black; dashed lines indicate suspected earthworks (redrawn from Kurz, S. 1998; Kurz, S. 2008, Figure 1; and Kurz, G. 2008, Figure 10) 82

Fig. 6: Plan of the structured landscape of the Glauberg showing the hillfort and a complex of earthworks, including the remains of Tumulus 1 at the northern end of a ditched passageway. The earthworks were interrupted to incorporate older mounds and urn graves at Enzheimer Wald. Ditches are shown in grey and embankments are shown in black (redrawn from Hansen and Pare 2008, Figure 1). 83

Ancestral Places
Phil Mason

Fig. 1: Dolenjska and Bela kraijina in the 1st millennium BC, showing major settlements (After Dular 1993: 103, fig. 1; with additions from Dular 1985: 31, fig. 12; ANSl 1975; Dular and Tecco Hvala 2007; drawn by Ildikó Pintér). 88

Fig. 2: Plan of the late prehistoric settlement complex at Vinji vrh (Source: Agencija RS za okolje; data adapted from A. Dular 1991, fig. 3; with addition of recent data; drawn by Ildikó Pintér). 89

Fig. 3: Plan of the late prehistoric settlement complex at Kučar pri Podzemlju (Source: Agencija RS za okolje; data adapted from J. Dular, Ciglenečki and A. Dular 1995, 8, fig. 2; with addition of recent data; drawn by Ildikó Pintér). 90

Plate 1: The Roman flat cremation cemetery at Mačkovec (Photo: Marko Pršina; Archive ZVKDS, CPA) 92

Fig. 4: Dolenjska and Bela kraijina in the 1st millennium AD, showing Roman and early medieval settlement and mortuary sites, including those with evidence of re-use of later prehistoric sites (After Dular 1993: 103, fig. 1; with additions from Dular 1985: 31, fig. 12; ANSl 1975; Dular and Tecco Hvala 2007; drawn by Ildikó Pintér). 93

Plate 2: The church of Sv. Helena on the large Early Iron Age barrow at Zemelj (Photo: Ildikó Pintér) 94

Recycling Pots, Places and Practices
Bernarda Županek and Irena Sivec

Fig. 1: The Podlipoglav area with sites mentioned in the text: 1 Molnik, 2 Podmolnik, 3 Marenček, 4 Sostro/sv. Lenart, 5 Zavoglje, 6 Besnica/Tomaž, 7 Češnjica, 8 Gradišče/Zagradišče, 9 Javor, 10 Ravno brdo, 11 Veliki Lipoglav/Roje, 12 Mali Lipoglav/Mrdiž, 13 Pance, 14. Magdalenska gora 98

Fig. 2: The Podlipoglav cemetery under excavation. 99

Fig. 3: Dated graves. 99

Fig. 4: Plan of the excavated cemetery; a suggested chronology of the burials. 100

Fig. 5: The grave goods in grave 2: combination of pot and cup in Latobican tradition, imported oil lamp, plate with red slip (imitation of sigillata ware), two vessel fragments. Photo Matevž Paternoster, MGML archive. 100

Fig. 6: Grave 2 during excavation. Photo Andrej Gaspari, MGML archive. 101

Fig. 7: Grave goods in grave 31. Photo Matevž Paternoster, MGML archive. 101

Secondary use of storage vessels and household pottery during the late middle Ages
Marta Caroscio

Fig. 1: "Iglesia de Santo Domingo" (Valencia, Spain), secondary use of kiln waste as vault fill. 107

Fig. 2: "Iglesia de Santo Domingo" (Valencia, Spain), secondary use of a jar locally produced and used (domestic waste) as vault fill 107

Fig. 3: "Iglesia de Santo Domingo" (Valencia, Spain), secondary use of an imported jar from Seville as vault fill. 108

The Reuse of Materials During the Medieval and Post-Medieval Periods
George Nash

Fig. 1: The centre of Rothwell showing the location of the former supermarket and the development area (after Gifford & Partners 2003). 112

Fig. 2: Plan of the Township of Rothwell dated to 1839 (the extent of the supermarket development is shown in red) 112

Fig. 3: The 1st Edition Ordnance Survey sheet of 1894 (the extent of the supermarket development is shown in red). 113

Fig. 4: Map showing the extent of the development and the building complexes and boundary plot walls that formed the assessment (after Gifford & Partners 2003). 113

Plate 1: A medieval timber-framed roof covered by a 19th century tiled roof within No. 43 Commercial Street. 115

Plates 2 and 3: The Commercial Street frontages and rear sections of Nos. 32-36 Commercial Street. 116

Plates 4 and 5: Moulded timber floral casement panel and plaster moulded ceiling. 117

Plates 6 and 7: Front and rear elevations of a 19th century terrace, north of West Parade which housed the medieval timbers. 118

Plates 8 and 9: Detail of the multi-phased building components and recycled materials used, building located to the rear of No. 30 Commercial Street. 119

Plates 10 and 11: The stable block with recycled cast-iron column and adjoining outbuildings. 120

Plates 12 and 13: The eastern [front] elevation of the former chapel and the recycled northern foundation wall that it once sat upon. 121

Plate 14: The former jail building located within the eastern part of Jail Yard and the approach to Jail Yard via Commercial Street (looking north). 122

Plate 15: The three-phased wall section that formed the western boundary wall or the rear plot. 123

Plate 16: Multi-phased sandstone wall forming the eastern rear wall of No. 50 Commercial Street. 124

Plate 17: Outbuilding/workshop with tiled roof, located to the rear of No. 50 Commercial Street. 124

Plate 18: Eastern section of a rebuilt N-S medieval burgage plot boundary wall containing dressed medieval stone, looking SW. 125

Plate 19: Reuse of medieval stone blocking in c. 1980, bonded with Portland cement. 126

Fig. 5: A mid 19th century east-west wall section constructed using a variety of materials. 126

Plate 20: Multi-phased wall section, located between Nos. 56 & 58 Commercial Street. 127

Plate 21: Small oven attached to the boundary wall section between Nos. 56 and 58 Commercial Street. 127

Plate 22: Medieval timbers stacked in a corner of the development area. 128

Plate 23: Mortice and tenon joint chiselled out of a roof collar. 128

Figures 6 to 9: A selection of recorded medieval timbers (ceiling joints) from a terrace north of West Parade; (© - carpenter's mark). 129

Plate 24: Carpenters/assembly marks on Timber 9. 130

Plate 25: Medieval dressed sandstone with chisel marks on the external face. 130

Plate 26: Post-medieval dressed stone, possible copping stone or mounded casement. 130

Plate 27: Probable 16th/17th century part vitrified brick. 131

Plate 28: Corner brick from the Cliff & Son factory in Leeds, late 19th century in date. 131

Contributors

Ian Armit
Division of Archaeological, Geographical and Environmental Sciences
University of Bradford, UK
I.Armit@bradford.ac.uk

Marta Caroscio
Universita degli Studi di Firenze, Dipartimento di Storia, Archeologia, Geografia, Arte e Spettacolo (SAGAS), Italy
martacaroscio@gmail.com

Davide Delfino
Land and Memory Institute (Mação-Portugal)
Group "Quaternary and Prehistory", Centre of Geosciences (uiD73 F.C.T.)
davdelfino@gmail.com

Roberta Robin Dods
Irving K. Barber School of Arts and Science
University of British Columbia Okanagan, Canada
robin.dods@ubc.ca

Andrew Dunwell
CFA Archaeology Ltd, UK
cfa@cfa-archaeology.co.uk

Dragoş Gheorghiu
Doctoral School
National University of Arts, Bucharest, Romania
gheorghiu_dragos@yahoo.com

Fraser Hunter
Scottish History and Archaeology Department
National Museums of Scotland
f.hunter@nms.ac.uk

Phil Mason
Institute for the Protection of Cultural Heritage of Slovenia
Centre for Preventive Archaeology
phil.mason@zvkds.si

Matthew L. Murray
Department of Sociology and Anthropology
University of Mississippi, USA
mlmurray@olemiss.edu

George Nash
Geosciences Centre of Coimbra University (Quaternary and Prehistory Group), Portugal;
Department of Archaeology & Anthropology,
University of Bristol, England
George.h.nash@hotmail.com

Per Nilsson
Department of Archaeology and Classical Studies
Stockholm University, Sweden
per.nilsson@ark.su.se

Lolita Nikolova
International Institute of Anthropology, USA
lnikol@iianthropology.org

Irena Sivec
Museum and galleries of Ljubljana, Slovenia
irena.sivec@mestnimuzej.si

Bernarda Županek,
Museum and galleries of Ljubljana, Slovenia
bernarda.zupanek@mestnimuzej.si

The Never Ending Journey:
Cycling and Recycling Seen through a Critical Assessment of the Taphonomic Process

Roberta Robin Dods

Abstract

In working through problems of interpretation of natural and cultural realms there is theoretical space for the expansion of the concept of taphonomy. Therefore this is an examination of taphonomy in both a literal and figurative sense. The loss of "information" is investigated from four perspectives. The traditional taphonomy based the physicality of material culture as defined by Popper's World One of existence/reality (1978) is familiar. But what of the seemingly fleeting and intangible domain of ideas, Popper's Worlds Two and Three? This chapter looks at a taphonomic interval for the ideational. Systemic contexts and filters of the natural and ideational/cultural systems will be considered and supported through specific examples. The impact of the loss of information on the interpretation of the world of the past, not the least of this being the recycling of its materials through the culture of archaeology, is thus discussed.

Plate 1: Item of war turned into a child's toy. An example of use and reuse from a Bedouin camp in Jordan. Image R.R. Dods 1990.

Contextualization: Understanding the Situated Self

Duden (1991) à la Hartley, observed in her book *The Woman Beneath The Skin: a Doctor's Patients in Eighteenth-Century Germany*, that the past is a different country even within the same cultural tradition. People there did things differently. None the less we want to assume that we share a commonality with those who came before us. This assumption, Duden (1991) contends, is false. We may have nothing in common with those of the past. However, this should not deter archaeologists from attempting theory building on the nature of the unknown cultural processes based on interpreting material culture. The unknown does not ultimately and irrevocably always mean the unknowable. I have been contemplating Duden's ideas about history for nearly a decade (Dods 2009; 2007a; 2007b; 2004; 2003; 2002; 2000) challenged by Feyerabend (1988; 1994), Lyotard (1987), Popper (1978; 1983) and most particularly various works and interviews of Ivan Illich (between 1971 and 1992), in particular his call to radical awareness. Illich's introduction to Uwe Poerksen and the concept of plastic words in a modular language published in German in 1988, which I was finally able to read in English in 1995, has stimulated considerable thought on what we know and what we think we know.

The acceptance of different points of view on the same "facts" and the ability to work on a synthesis of Traditional (cultural and ecological) Knowledge and Western Scientific Knowledge (Dods 2004) forms of inquiry challenge our ability to work out a "story" that remains as true as possible to the material and meaning of archaeological remains. Here narrative archaeology (Joyce 2002) and agency theory (Dobres and Robb 2000) can contribute to reconciliation of these ways of interpreting "data". This process can be somewhat straight forward where we have historical, proto-historical, ethnographic, and/or direct interview materials to work from. However, analogues (same function, different origins), homologues (same origins, different functions), and metaphorical models become tenuous when moving beyond the boundary of written documents and direct oral tradition into the realm of deep time and different traditions. Each suffers from the problem of being thrice removed - removed by time, removed by space, and removed by cultural tradition (diachronic removal/synchronic removal/ideational removal). The choice of interpretive models has to be carefully considered if we do not want to come to the essentially political question: *"Whose facts?"* Indeed, in well considered models of interpretation all potential results would have a place in the inquiry and its outcome.

Consider contextualisation of our work. The context of what is being done is the context of archaeology, and essentially in the cultural analysis aspects, anthropology. It consists of both science and social science and we could add the humanities if we include the methodologies of the historian for literate societies. In the academy each of these inquiries has pattern and structure. However, that which is being observed happened in the context of the life of an individual, of the relationships of specific groups (kin and non-kin) (David and Lourandos 1998), of a society, of a culture other than the one from which the observations are being made. Regardless such places being of the literate tradition or pre-literate-oral tradition in record keeping, such social/cultural places can never be visited or observed directly in their dynamic actuality. They can never be completely, exactly re-tested through archaeology and the ability to replicate the test exactly is a central tenet of the Western scientific model. What we do is not the re-testable experiment but rather sampling of a specific "universe" - a form of inquiry that is founded in statistics. Effectively, it comes to probability statements on specific cultural remains. Such sampling relies on a universe of material remains that are 'facts' from which probabilities are constructed on selected "members" of that universe. Their complete distribution can never be known, only more or less approximated. Distribution is ruled by taphonomic processes.[1]

The worlds of the past had their own patterns and structures - each operating within the constructs of particular societal philosophies. They had their own contexts. Patterns can be discerned, more or less. So although we can never reconsider the complete universe of any past society we can, to some indefinable extent, see context as pattern and structure. It is there to be un-covered, dis-covered, re-constituted, re-constructed. The details are obscured by distance and time and ultimately by our methodologies, our skill in methodological application, and then by our informed imagination in understanding what we have observed as 'data'. The peoples of these other places had the knowledge that made them members of their own groups. The archaeologist is attempting to gain membership in one of these groups. But unlike the enculturation process of the anthropologist in fieldwork in the development of an intersubjective space, there are no teachers to bring him or her through intellectual 'childhood' to an 'adult' understanding within the subject society. The intersubjective relationship of participant observer and informant is turned into a one-sided conversation, a conversation with the self. Although informed by fragmentary material cultural remains in a specific environment, it is our reflexivity that allows us to come to understand the *data*. This is the 'culture shock' of the archaeologist (Dods 2007a). This is Duden's (1991) angst about history.

Then there is the context of who benefits from the research - for whom is it being done? Is this for the self-satisfaction of the researcher, for the demands of the academy, for the interest of members of the researcher's own society or culture? Or is it for the society of the "others", if they are yet represented by an identified or identifiable culture group in the present? Is it accessible to them or their descendants or the public in general; does it resonate with meaning or will it be viewed as an archaeologist making a living? Is there political content and for what purpose? These are questions that need to be kept in mind when we do our work as they drive the agenda regardless of our recognition or lack of recognition of them in an overt, conscious way.

Popper's Worlds: The Archaeologist in Wonderland

I want to begin this consideration of taphonomy by briefly outlining the worlds of knowledge Popper proposed in a lecture in 1978. They succinctly outline where we are in archaeology when we try to understand the nature of our data.

World 1: Physical objects. It is our physical universe. There is an actual truth here in as much as we can quantify/describe this world with the caveat that systems of what is suitable for quantification vary from culture to culture. Additionally, we all do not perceive *things* in exactly the same way as some of our understanding is tacit (Polanyi 1966). Thus its representation can be idiosyncratic, culturally dependent and difficult to transfer generation to generation. This leads us to World 2.

---

[1] A basket may remain in record in a dry environment. As a trade item into a specific wet environment its use there will disappear into the memory of time – as though it never was.

World 2: Thought/subjective experience. It can describe World 1 and interact with it from the subjective experience (hopefully informed by reflexivity) based on our cognition (a physicality tempered by cultural constructs) and experiences of the physical world and our subjective response. It can draw on it to create from raw materials items for World 3. In computer parlance it is wetware. Here the tacit emerges and so too the explicit and the word becomes manifest in World 3. The ability to display and share thought and to transfer it through oral tradition as well as documents may or may not stand the test of time in World 1. There is the tyranny of taphonomy in the mechanical, chemical, and biological actions of this interval that take not only things but ideas too!

World 3: Products of the human mind. In literate societies **objective, abstract** "products" of the human mind (World 2) in becoming substantive can be "explained" in forms of non-oral communication. They can become material objects composed from items in World 1. BUT through time they may become impossible to understand nonetheless (think of the Indus Valley script yet to be deciphered or metaphorical statements that seem senseless to us today). We have wetware (actions of the brain in World 2) in software (written instructions of World 3) to run hardware (computers also World 3). What if we as archaeologists of the future find only the hardware? Knowledge created in World 2 become hardware artifacts in World 3. They speak to technology but do they speak to deep meaning? In oral tradition societies (pre-history) there is the missing instruction manual beyond the cultural concept of tradition – "we do it this way because we always do it this way". Cut out the "tongue" of a society and you destroy the relationship between World 2 and World 3.

Taphonomy in Three Worlds

In working through problems of interpretation of natural and cultural realms it became clear to me that an expansion of the concept of taphonomy, borrowed so usefully from the natural sciences, was necessary. The critical question became:

Can we come to understand the biases presented in the record and our role as archaeologists in the formation of some of these biases?

There is a progression to the loss of information:

1. Consider what could be and is chosen from the natural system for use by humans. This speaks to cultural decisions of use/do not use. This is the articulation point between nature and culture and it is within the traditional taphonomy in the sense of its use by palaeontologists. This is most particularly so if humans are seen as having culture as species specific behavioural adaptation. How much do we get to see (Point 3, below)?

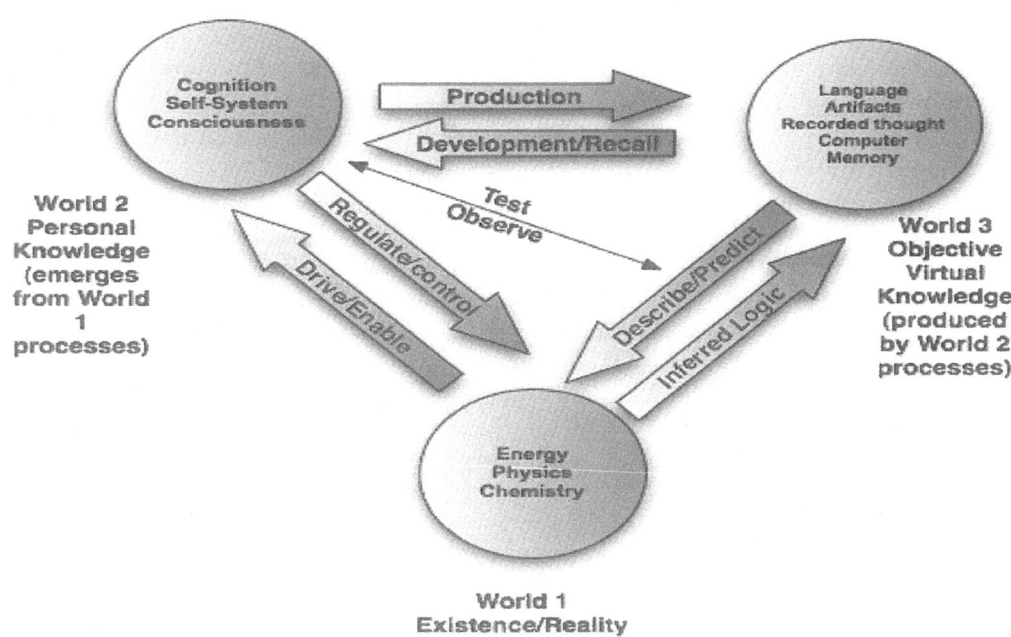

Fig. 1: *Karl Popper's Three Worlds of Knowledge. (http://www.knowledgejump.com/knowledge/popper.html)*

2. There is the "site" itself and the distribution of material culture in the broadest sense. Retrieval of any of this is but a sample of a past reality. Our ideas on what constitute a site, where its boundaries exist, how we define its catchment (Point 1, above), and our sampling decisions are important here.
3. What happens to the lost, discarded, abandoned – the "place" (frequently middens but not exclusively such, for example kill/butchery sites (Saunders and Daeschler 1994))? We have conventionally centred our attention on the processes of the taphonomic interval in such locations.

This Taphonomic interval is

- tinged by the cultural events of the living system preceding entry into the non-use realm, however this happens;
- shaped by the cultural constraints/dictates on disposal through discard,[2] loss, abandonment;
- formed by the 'natural' events during deposition – biological, chemical, mechanical;
- structured by post-deposition cultural events, one of which is called archaeology.

Including the archaeologist in the taphonomic interval gives symmetry to the investigation. This symmetry brings the culture(s) of the past into articulation with the culture of archaeology through the agency of humans. Understanding this complexity is needed if we are to deal with both the natural and cultural worlds in a way that we can approach the nature and extent of information loss between the past and its investigation in the present.

It can be challenging to present a dynamic four dimensional process (Baltensweiler and Fischlin 1987; Krebs 1985; Loeschcke 1987; Turner and Gardner 1991a; 1991b; Wheeler 1990; Wheeler 1954) in a two dimensional format. Following on the outline of the flow model in Fig. 2, I think Fig. 3 summarizes, somewhat, the concept and spirit of information loss in systems (Clarke 1968; Schultz and Zwolfer 1987) that integrate the natural and cultural realms in which humans live. It incorporates both the synchronic and diachronic in the structure of the model as a diagram because of the representation of the four dimensions. As such, the diminishing volume of the 'black boxes' and their "filters" represents the diminishing volume of data, both concrete and ideational, with the elapse of time.

---
[2] Northern Algonquian traditionally returned fish bones to rivers and lakes. Such bones are in low in absolute numbers in middens although historically fish are known to be in the diet. The standard archaeological explanation has been low pH but small mammals are in similar contexts. My work shows that pH values can vary considerably but fish remains are relatively static in percentage representation when size of bone and fragment is considered.

In this instance the boxes operate both as a convenient way to construct the diagram and as a representation of the 'black box' process in systems analysis - the hidden, the unknown, for which you may know or learn the input and the output but perhaps never the complete process (systems operation) between the two. On another level the box model is part of our history if we consider the Wheeler Box (an excavation technique to facilitate the retrieval of more information). Here the individual contexts, or units of information seen as "boxes" only because of the constraints of graphics, can be 'opened' and, much like Russian dolls, can be studied individually in more detail in a flow diagram (or have alternative flow diagrams 'tried out') for specific processes at various levels of integration. The separate and unique problems of each context thus can be expanded in additional diagrams befitting each set of circumstances. Flow diagrams function well in such detailed cases as they give us the opportunity to play out different scenarios – develop different narratives and even flag areas of lacuna that we may attempt to address. Indeed, preliminary flow charts in the field can help frame alternative and additional approaches to retrieval of data. What becomes evident is that this process of 'opening' can become almost endless as finer and finer detail is sought.

The filters affect material culture through taphonomy of the concrete (things composed from items in Popper's World 1). But loss is also to be found in the interpretation of past cultures (Worlds 2 and 3). In the loss of the concrete (for example NSC:B as in a midden) the ideational context of the archaeological culture (CSC:A) is diminished as well. A case in point could be the surface decoration of a bone harpoon point that is leached away by the action of low pH percolating water. This may be considered to be incidental to the central processes of bone loss since this is the result of the actions of the natural filters central to traditional application of the concept of taphonomy (mechanical, chemical, and/or biological destructive processes). However, three types of information are lost: information on the object as material from the natural world (bone), information on the object as manufactured object (tool), and information on the object as symbol/symbolic item (design). The latter two are of the realm of culture but so is the first in that it indicates cultural choices of selection from the natural realm.

The two natural contexts, NSC-A and NSC-B, can be investigated best under the rubric of 'traditional taphonomy'. However, the term taphonomy as first posited refers best to the actions in NSC:B. Certainly it works on a metaphorical level but I think it has validity beyond that into the operational level as well.

In discussing Taphonomy I the specific illustrative environment used here is the circumpolar northern hemisphere generally and boreal forests specifically

(Barbour and Billings 1988; Collins and Wallace 1990; Delcourt and Delcourt 1981, 1987; Hills 1959a, 1959b, 1976; Laford 1958; MacDonald 1987a, 1987b, 1984; Potzger 1946; Reichle 1973; Shugart, *et al.* 1992; Tamm 1976). However, the concepts can be generalized to other environments and Krebs's 1985 breakdown of the biosphere in effect outlines the various levels of analysis for the archaeologist as well as the ecologist. Some of the questions we can attempt to answer are:

- What can one say about properties, forces, flow pathways, interactions?
- Can one talk of fluctuations, perturbations?
- Can one use the natural filters such as seasonality and periodicity as part of a catchment analysis and thus develop realistic models of fundamental and realised niches used by humans?

The fractal dimension at which the pattern emerges from the background noise may be such that, for example, fundamental niches may be theoretically constructed but their actualisation in realised niches, as they pertain to humans, may not be completely discernible. As an example let us look at boreal fire regimes (Dods 2007a). Fluctuations can give insight into the natural cycles of the boreal forest but in studying perturbations there may be problems in the use of pollen and charcoal to distinguish wild fires from fires of anthropogenic form – the form that informs us on human interaction in the natural systemic context. Further, there is the problem of defining the frequency, extent, and intensity of all fires that occurred in deep time. Such data could lead to a model for pyrotechnological economic landscape development and an explanation of how a system "worked". However, the absolute, empirical evidence for the process can yet eludes us. This has to do with the loss of detail in the information about the past environments. Some of this will be loss of biological materials such as the trees themselves, while some of it will be simply that we cannot go back and do the testing while certain events are taking place. In this latter instance, the example would be the ability to study pre-fire, fire, and post-fire conditions for that long past event and the actual choices that may have been made in the use of pyrotechnology to shape cultural/domesticated landscapes.

Plants and animals may offer us excellent information through the use of biological uniformitarianism. Figures may be developed for populations of animals, birds, and fish. This can be a rather questionable approach if we use modern studies as completely analogues without considering the changes that humans have made to the natural world in the last 10,000 or so years – for example the low population numbers today of the North Atlantic Cod or the Bison of the Great Plains of North America. The fidelity of theoretical reconstruction is the problem. It may be narrative in its actual approach if the tyranny of time and circumstances is not considered. In North America some of the contexts for failure in understanding the past through the use of analogy are based in the 'new' economic objectives applied to this modern day 'New' World. This means that estimated figures of population size for humans as well as all the other life forms (even trees) from this environment may not offer us an adequate picture of the dynamics of American prehistory. There has been an ideational manipulation of our imagination based on our Old World origins (Dods 2002). Our best hope to develop figures for prehistoric realities may rest with the early historic numbers in association with modern estimations worked through to plausible reconstructions informed by traditional indigenous concepts of environmental relationships.

To push this example, as I have noted elsewhere (Dods 2007b), the forests of the past were managed forests but with some different outcomes. Why I make this distinction of "some different outcomes" is that the management of the temperate forest for the purposes of agriculture (whether you call it slash and burn or swidden agriculture) has little to recommend it as being different from the clearing of temperate and boreal forests for cellulose and timber products. The outcome is the same - loss of species diversity in the farmed forests. The management of the forests for **the living in** is qualitatively (forest structure) and quantitatively (species diversity and numbers) different. Theoretical catchment for a specific population can be suggested from understanding the structure of boreal forests and the environmental needs of the species in these environments, creating flow statements of functional integration and limiting factors. To do this one has to be clear on the levels of integration in biological systems and the exact ones that are the focus of specific studies. Here can be the tyranny of the narrative based on selective and/or incomplete data – taphonomy of information in the hands of the researcher.

With Taphonomy III in Natural Systemic Context: B (NSC:B), the refuse of a past culture or cultures, can be affected by cultural filters in midden development and discard regimes. These are contingent on site size, site type (camp, village, etc.), duration of site use, amount, type, and choice of refuse discarded in relation to the site as a whole. Depositional history will affect the preservation potential of a specific deposit. The fact that depositional history can be very complex is attested to by the debate on the interpretation of re-deposited material from the Old Crow Basin in the Yukon (Morlan 1980). To paraphrase Behrensmeyer's observations on fossilised material (1978: 150), the survival probability of any object depends on the intensity and rate of a host of interacting destructive processes and the prospect for permanent burial before total destruction has occurred. For archaeological

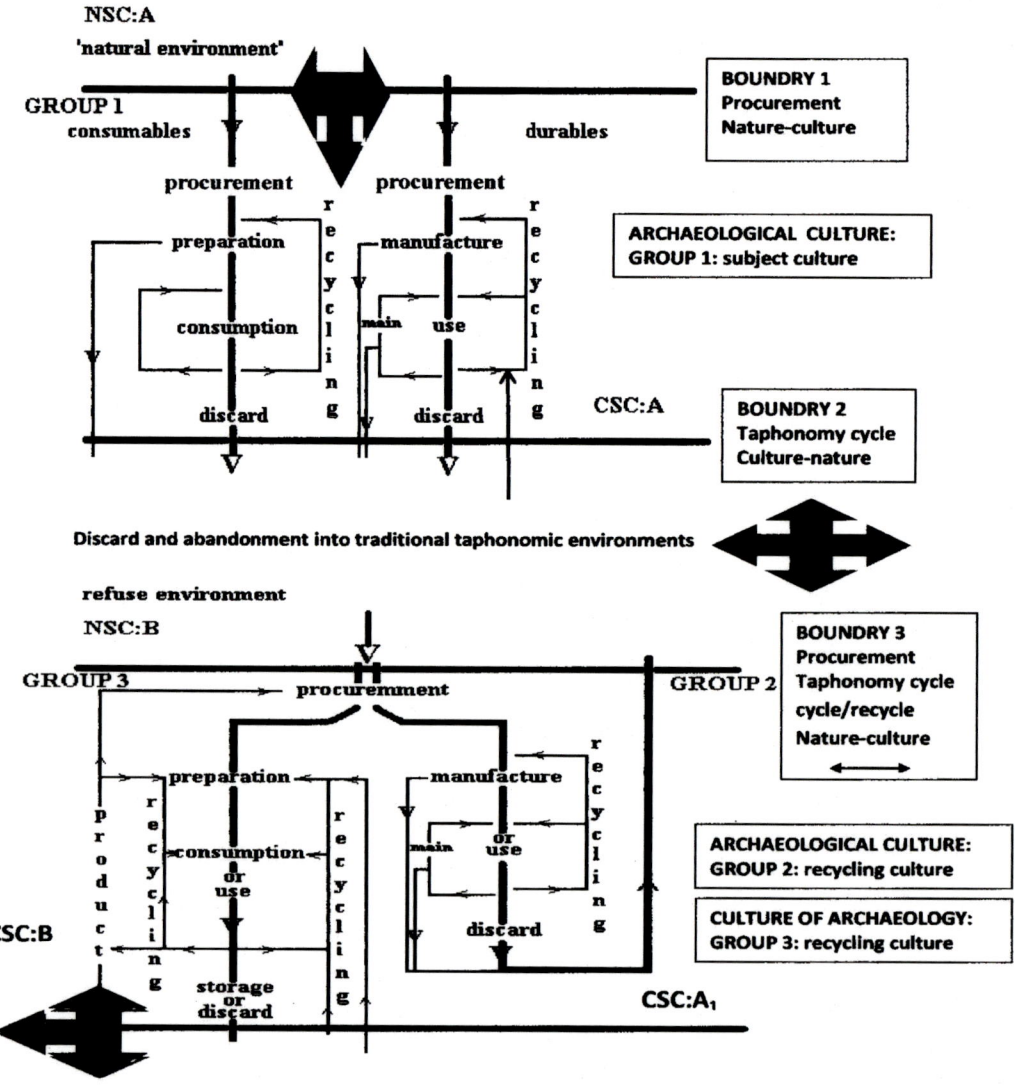

Fig. 2: *Flow model for archaeological materials adapted and expanded from Schiffer (1972: 158-159), Lange and Rydberg (1972), and Clarke (1968: 36). Note the material remains of subject culture (1) in systemic contexts of GROUP 2 and GROUP 3 (the archaeologists). Recycle is recycle regardless!*

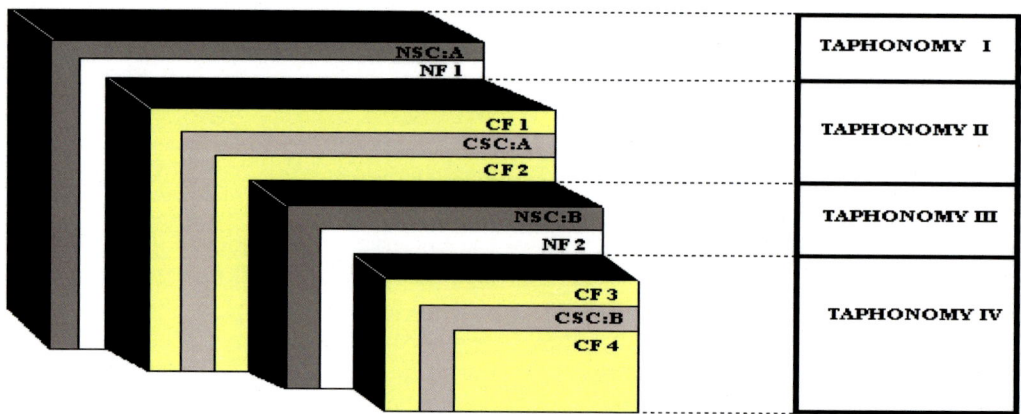

Fig. 3: *Model of the linkages/relationships of taphonomies I through IV. The main feature is diminishment of information. Three dimensions of space (in the diagram represented by the cube) and the dimension of time (represented by the movement into the visual foreground of the model on the page) are depicted.*

Chart 1: *CONTEXTS AND THEIR MEANING FOR FIGS 2 and 3*

SYSTEMIC CONTEXTS:

Defined for our purposes as either Natural (NSC) or Cultural (CSC)

The context in which transformational processes take place:

1. NSC:A: ecosystems, habitats, natural environments...
2. NSC:B: archaeological discard environments (e.g. middens)...
3. CSC:A: archaeological *subject* culture
4. CSC:B: recycling culture(s)
   - Group 2, an archaeological culture using items from Group 1
   - Group 3, the culture of archaeology

FILTERS:

1. NATURAL FILTER (NF):
   The actual processes that cause loss of information within a natural systemic context. For example, low pH in midden soils will cause bone disintegration. Thus pH is a chemical filter in Natural Systemic Context B.

2. CULTURAL FILTER (CF):
   The actual processes that cause loss of information within a cultural systemic context such as the loss of ideational information in oral tradition societies that are removed by both time and space and have no direct link to us in the 'now'. For example information on food taboos (what is eaten / not eaten), conventions of food preparation, hunting myths.... and eventually the culture of archaeology and its theories and methodologies.

TAPHONOMIES AND SOME OF THEIR CONTEXTUAL QUESTIONS

1. TAPHONOMY I: NATURAL SYSTEMIC CONTEXT A
   What can we reconstruct of the original biotic community?
   Can we speak of populations or only individuals?
   Can we discuss human niches?
   What level of functional integration can we achieve?
   What about properties, forces, flow pathways, interactions?
   Can we talk of fluctuations, perturbations?

2. TAPHONOMY II: CULTURAL SYSTEMIC CONTEXT A
   What can we say about the cultural system of
   - acquisition?
   - use/consumption?
   - recycling/discard?
   - discard itself?

3. TAPHONOMY III: NATURAL SYSTEMIC CONTEXT B
   What do we know of the depositional / dispositional history?
   What can we say about the processes that are
   - biological?
   - chemical?
   - mechanical?

4. TAPHONOMY IV: CULTURAL SYSTEMIC CONTEXT B
   What are the critical practices (choices) in:
   - acquisition?
   - use/consumption?
   - recycling/discard?
   - interpretation?

materials that are biological the same parameters pertain as for paleontological materials. However, because of the nature of cultural activity, some of the processes of destruction have broader consequence on the understanding of culture. So there is a need to develop an understanding of natural filters (NF2), the diagenesis that consists of the chemical, biological, and mechanical activities that diminish organic systems from tissues to cells to molecules until the 'star dust' of Hawking's description (1988) is re-achieved. This is the loss of the integrative information of the concrete (see in part: Binford and Bertram 1977; Casteel 1971; Gifford 1978, 1980; Hill 1976; Lyman 1985, 1994; Lyon 1970; Morlan 1980; Payne 1972a, 1972b; Piepenbrink 1986; Schiffer 1972, 1987; Speth 1983; Williams 1992).

Cultural decisions have repercussions on the process in the natural system and outcomes for the deposited materials associated with cultural activities. Within this context, culture is indeed part of nature and they have similar processes in the mechanical, biological, and chemical diminishment of the original items from time of acquisition until discard and then recovery. This is the loss of information from these materials. For example, a specific society may have rules about how food is cooked. A butchered fowl's bones end up in a soup pot in a liquid environment that is acidic (say lemon was added) to be boiled for hours. Calcium will be leached from the bones into the resulting broth - all good for human nutrition. However, it is not so good for the bones as items of information for the archaeologist unless cooking information is the only objective of the research. So this is a cultural decision on how poultry soup is made but it also has repercussions for the physical preservation of biological materials, the bones of a bird in this case. Decomposition on subaerial surfaces "...can be viewed as part of the normal process of nutrient recycling..." (Behrensmeyer 1978: 150) but this offers little comfort to the archaeologist when the survival probability of an artifact in an archaeological context depends on the intensity and rate of a host of interacting destructive processes and the prospect for permanent burial before total destruction has occurred.

A combination of processes can effectively narrow the data to uselessness. However, because of the nature of cultural activity, some of these factors have broader implications in the destruction of materials in the archaeological context then would apply for materials from paleontological sites. Mechanical destruction will now include human activities such as the butchering of animals or the fragmentation of many items small and large, some of it ancient and some of it modern. For examples we may consider the 'killing' of pots as part of ancient ideational regimes, the destruction of the Bamyan Buddha statues in Afghanistan because of modern religious concepts, or, from our modern activities, the acid rain destruction of whole sites (think of Angkor Wat in Cambodia or the walls of hieroglyphs in Egypt). Therefore to the traditional categories of destructive regimes of hot/cold, thawing/freezing, moisture/desiccation, transportation by natural agencies, and the effect of wind exfoliation can be added all the alteration agencies directed toward archaeological materials by humans as a result of their culturally prescribed concepts of acquisition, preparation, use, consumption, discard, and recycling. What comes to mind are the myriad of items from which housing and garments, weapons and tools, binding and storage, decoration and medicine, warmth and symbol are composed.

Taphonomy II and Taphonomy IV specifically address the issue of cultural systemic contexts (CSC:A; CSC:B) and their associated cultural filters. Material items move through these systemic contexts. The forces diminishing information are not only the biological, chemical, and mechanical but also the forces at play in ideational systems. Ideational forces can cause changes to occur in the physical world. For example an animal lives or dies in its encounter with the hunter perhaps on the basis of food taboos or some edifice is constructed because of concern for the proper recognition of a deity or a universal force. Through time, ideational constructs also change. This change in the ideational may change the boundaries, content, and shape of the physical / metaphysical world(s) as noted with the Bamyan example. It may alter the tone of discourse; change world views. The process may be subtle and / or it may be cataclysmic. Many of the forces that cause change, and the direction and outcome of change are difficult to determine from physical remains. Most of the remains archaeologists work with, in particular in small scale social settings, do not inform on the non-physical construct, do not tell us 'why' and frequently do not even let us know 'how'.

Cultural Systemic Context: A (CSC:A), with its associated filters of CF1 and CF2, is the context through which the processes within an archaeological culture can be explored. In the model here these are called Group 1 and Group 2 in the attempt to incorporate the possibility that discarded items from Group 1 may be recycled by a subsequent archaeologically defined culture (Lange and Rydberg 1972)(see Plates 2 and 3); indeed this may be expanded to include traded-in items or even be used to follow traded-out items. Thus Group 1 is the specific target group of the archaeological investigation while Group 2 is a more general category and includes all other cultures that interacted with the target culture or acted upon the 'refuse' of the target culture, excluding the culture of archaeology, which is dealt with under Taphonomy IV.

The processes occurring here can be viewed from a cultural perspective as well as a biological - ecological perspective. There are two very basic things that all societies must work through: the process of energy production (tools); and the process of energy maintenance (food, clothing, shelter). This may seem so very functionalist but there are things that need to be done to stay alive! From primary procurement onward there is the potential for the production of varying types of refuse depending on the use decisions made at specific points. These types of refuse have their culturally specific patterns of discard some, as in the footnote example of the fish bones, dictated by concepts of the metaphysical.

The progress of consumables and durables can be examined within the same model. Durables are 'consumed'[3] just not in the way of, say, foodstuffs. They have a lasting quality to them. An excellent example is pottery. Pottery can only come into existence by the acquisition of materials from the 'natural' world (clays, minerals, etc...) being processed in a manufacturing model that includes other items from the natural world such as fuel for firing. It works under certain cultural constraints: division of labour, specific pottery forms for specific functions, ownership of designs, etc... It is then used, traded, abandoned, broken, discarded, and even 'killed'...recycled as grit for temper for future pottery and on to the discard environment and then the hands of the archaeologist who studies it for all these aspects and more. It can be followed from acquisition and transport through to processing and storage then on to use (reuse) and then to eventual discarding of what the culture considers the detritus (including faeces and excreta) of all stages of preparation and use. For models of this process see Schiffer (1972: 162); for a broad discussion see Clarke (1968); for site formation processes see Gifford (1978: 80); and for the biological aspects of this process see the diagram found in Delany (1982: 99). The constituents of both consumables and durables will move from the natural realm to the cultural realm because of **concepts of use**. Procurement, processing, reprocessing, use, discard all have culturally specific operational locales. Indeed, sometimes 'waste' itself is recycled within a culture.

Some insight on spatial distributions and their cultural content came from my ethnographic observations during fieldwork in Botswana in 1978 (Dods 1979). I came across what I considered an interesting example of the allocation of food resources on the basis of gender and status. In the traditional Tswana view the head of hunted animals (and incidentally the very few slaughtered cattle since traditionally cattle were not grown for meat specifically) went to the Headman along with the kidneys and liver. It was his obligation to give the kidneys and liver to widows and orphans. The head was his and he could share it with other men if he wanted. The Headman would sit in what we would consider his front yard. Actually the low walled space in front of his house, a rondavel (Plate 4), is a private space that he can make public if he so chooses. Passing men (never women or children) could be (making public space), or could not be (now making private space) recognised by the Headman. On recognition the passing male would/could be invited to share the eating of the head and of course the brain. The fragments of bone from skulls were in high concentration in a rectangular area enclosed by the small wall in front of rondavel found in archaeological contexts. Thus concentrations of skull fragments in specific locations in an archaeological context, in southern African Iron Age sites could be interpreted as a prehistoric record of this social activity so recently observed. This example has deeper nuances of gender and nutrition, food as status items, and group sharing as forms of solidarity than explored here; nevertheless you, the reader, are challenged to consider and imagine.

Therefore, with food one may ask what was available as food, an inquiry that flows directly from the work on NSC:A. From available resources what evidence do we have for food choices? What food appears to have been ignored? In other words: 'What was consumed or not consumed as food?' Or, as one Algonquian informant observed: "We don't eat that (an amphibian), its yucky". What is the impact on diet? (Barnes 1976; Chaney and Ross 1971; Dennell 1979; Farkas 1979; Hegsted 1976; Høygaard 1941; Lyman 1979; Price 1985; Wing and Brown 1979) To what extent can operational chains for consumables be defined? What 'products' of these processes or 'links' in the chain can one distinguish in the archaeological record? Can one speculate on the ideational categories, as seen in *Le cru et le cuit* (Lévi-Strauss 1964), of the natural world from the fragments of such processes? What is the role of analogy as seen in the direct historical approach or the application of what can be termed cultural uniformitarianism, in the reconstruction of the past? What categories of information are lost? And then what categories of information are retrievable? Can one deduce the ideational systems that dictated the choices seen or are the frames of reference so different that such speculation results in artful, albeit unintentional, deception? Can we even recognize deception?

There are at least two sets of ideological constructs interfacing in the archaeological process. There is the culture of the past - the archaeological culture (the 'object' of investigation, the 'subject' of the report) and there is the culture of the present - the culture of archaeology (the 'investigator' and the 'reporter'). As I noted earlier, there is the problem of the development of the intersubjective when the culture of archaeology acts on the archaeological culture.

---
[3] For example, breakage and discard can happen, retooling/retouch with debitage resulting may occur, items that go out of style can be redesigned or discarded or moved into other uses by another group – based on age, class, gender, etc...

Archaeologists as Actors in Taphonomic Interval

Cultural Systemic Context B (CSC:B) and its filters, CF3, and CF4 like CSC:A deals with acquisition, use / consumption and discard cycles. However, here the culture being examined is the culture of archaeology itself (Group 3). Archaeological parameters encoded by existing institutions and practices affect the procurement: survey and selection of sites for excavation; decisions on the extent of the investigation; retrieval techniques to be employed, concepts of what constitute data; recording of the procedures. Preparation and 'consumption' (cataloguing; analysis; literature support; *etc.*), discard (storage or actual discard), and synthesis of the data (contribution to the discipline and to knowledge in general) continue the archaeological process. To sum up, cultural filters determine how the researcher discriminates in the retrieval process, and in the final analysis (no pun intended) how the material will be interpreted, used and discarded. The ideational component engaged in this, at times, as nebulous as that for the target culture, is the culture of archaeology. Perhaps this section of the model represents the one component most removed from the traditional use of the term 'taphonomy' although there are still physical 'things' that can happen to the archaeological remains. For example, material culture items may be damaged in the excavation; they may be destroyed in the analysis process such as with the process of radiocarbon dating. They may be lost in transit, or improperly stored. Any of a myriad of things can happen, purposely or inadvertently, to the physical structure of the "finds". In this sense the taphonomic process discussed in Taphonomy II is continued.

The loss of information (taphonomy used in a metaphorical sense) may be greater here than anywhere else in the model and this has little to do with the bumps and knocks that physically diminish the artefact. Rather it has to do with the skill of the interpreter and the intellectual, ideological, and/or political atmosphere of the environment in which he or she works.

There are certain points of disarticulation between the community of the past and the community of the present. These are the discontinuities of space / time normally thought of as being between peoples of the present and peoples of the past. But there are also those discontinuities that transcend physical displacement. Different models of knowing (TEK and WSK) and the crisis of narrative experienced by aboriginal peoples I have noted elsewhere (Dods 2004) speak to the point here. The economic and political discontinuities between the present representatives of the prehistoric peoples and the archaeologist as representative of a dominant sector of society have also been noted. I have also commented on the economic and political constraints of doing archaeology in the boreal north and this is not a unique experience. The choices made by the archaeologist in this atmosphere will affect the 'story' being **_produced_**. Now, if we want to diminish these influences on the 'story' then we have to recognise that there are such factors and that these factors overtly and covertly impinge on the process of interpretation.

One of the important problems is the nature of the questions being asked. We seldom regard questions

Plate 2: *Recycling discarded items witin a culture (potential Group 1 returned to Group 1). (Hudson 2009)*

Plate 3: *Dumpster diving. Recycling of food from discard area. Julia Golomb (left) and Alison Abreu-Garcia, both of Somerville, mine a dumpster of a metro-area grocery store. (Baker 2009 with Globe photo by Gretchen Ertl)*

Plate 4: *Mahalapye, Botswana traditional house with walled 'front yard'. (http://en.wikipedia.org/wiki/File:Mahalapye_traditional_house_cropped.jpg accessed 22.01.2017)*

as structuring answers but they can be directive generating answers within the context of the definition of what constitutes science and the practice thereof (e.g Bartley 1987; Downing 1991; Dunnell 1989; Embree 1989; Feyerabend 1994; Hanen, *et al.* 1980; Heal and Grime 1991 Osler 1980; Peters 1991; Popper 1983; Smith 1976). Additionally, a particular theory or methodology may catch our fancy, to be used with apparent success then continue in use without any returning for 'catch-up' on its subsequent development within the parent discipline. Loss of immediacy is perpetuated and loss of information results. As well information that comes from the development of new methodologies may not result if we ignore the disciplines from which we borrowed theories and methodologies. They are, in themselves, dynamic forms of investigation although they too may have stagnation problems as well.

In a wider perspective we may ask the question: **What is the form of the investigation and is it driven** by the quest for knowledge and/or learning as distinguished by Lyotard (1987)?[4]

Here the challenge is to see that there is an ideational taphonomy and it is shaped by our theoretical and methodological choices. Further, it was shaped by other cultural decisions – those made by the members of the subject culture in the past. Our insights and answers will drive the agenda of future theory and methodological applications.

Consequently ideational data encompass information for interpretation since humans live within culturally created landscapes of meaning (for example *built environments* from gardens to temples, tool sheds to palaces, concretely represent the ideational). Cultural information encompasses the environmental/natural/metaphysical in various, morphing combinations (Bielawski 1989; Bishop 1970; Cooper 1939; Gould and Cohen 1994; Hickerson 1967; Lévi-Strauss 1964; Rogers and Black 1976). Aspects of cultural information become manifest when they are superimposed on elements of and from the natural realm and then become embedded in the manufactured[5] item (Popper's Worlds). For example a bone projectile point includes not only the material from the natural realm from which it is made but the cultural realm as well. It is *manu-fact-ured* to a specific 'type'; because of this internal consistency of manufacture archaeologists could develop the use of the theory and methodology of typology to sort and define based on *facts* from the past. The tool may have been used for a precise purpose. The acquisition of the bone material as a resource may have been incidental to the hunt for food or it may have had specificity for the production of non-food items made from animal fibres of skin, sinew, bone, fat, and/or blood. Even more deeply embedded may be the fact that this bone may be from a specific animal and even a specific element in the skeleton of the animal (e.g. the ulna) (Bonnichsen and Will 1980). Further it may have been ritually required to make this specific point for a specific literal or metaphorical/metaphysical activity.

Such are the sometimes seemingly capricious outcomes of taphonomy that much comes to us inferred rather than in its direct manifestation. In many areas of the world ethnographic evidence tells us that the manufacture of twine was women's work, and that twine was used for diverse applications. Therefore the additional example of the manufacture of twine for assorted purposes is edifying. Indian "string" made from cedar, or from basswood in some southern North American areas (Vennum 1988: 83; 86), is easily biodegradable in the soils of moist environments. Knowledge in the historic period may come to us through ethnographic reports (Vennum 1988; Densmore 1929). However, information on twine in the prehistoric context may come to us only indirectly as basket formed items[6] or through the external decoration of pottery such as various wares that have cord-wrapped stick or cord-wrapped paddle impressions on the body, shoulder or neck area of pots (Dawson 1976; Dawson 1975; Lugenbeal 1978; McLeod 1978; Wright 1968b; Wright 1967a; Hurley 1979). There is then, embedded in the object from the past specific elements of information, some direct and some indirect.

Those elements from the natural world, if they survive the taphonomic interval, like the bone in the bone point or the clay in the pottery vessel, yield their information to us in a rather direct fashion through the application of specific laboratory methods. Here we can think of the survival of the pottery but the loss of the twine itself, although in this we have information on the twine through the imprint it made on the plastic medium of the paste that eventually became the non-plastic medium of the fired pot. This fits

---

[4] a. Knowledge:
..cannot be reduced to science, nor even to learning...what is meant... is not a set of denotative statements; far from it. It also includes notions of "know-how," "knowing how to live," "how to listen" [savoir-faire, savoir-vivre, savoir-écouter], etc. Knowledge, then is a question of competence that goes beyond the simple determination and application of the criterion of truth, extending to the determination and application of criteria of efficiency (technological qualification), of justice and/or happiness (ethical wisdom), of the beauty of a sound or color (auditory and visual sensibility), etc....it coincides with an extensive array of competence-building measures and is the only form embodied in a subject constituted by the various areas of competence composing it ....The consensus that permits such knowledge to be circumscribed and makes it possible to distinguish one who knows from one who doesn't (the foreigner, the child) is what constitutes the culture of a people (1987: 78-79).
b. Learning:
Learning is a set of statements that, to the exclusion of all other statements, denote or describe objects and may be declared true or false. Science is a subset of learning. It is also composed of denotative statements, but imposes two supplementary conditions on their acceptability: the objects to which they refer must be available for repeated access, in other words, they must be accessible in explicit conditions for observation; and it must be possible to decide whether or not a given statement pertains to the language judged relevant by the experts (1987: 78).
c. The relationship between the two:
...narrative knowledge does not give priority to the question of its own legitimation and that it certifies itself in the pragmatics of its own transmission without having recourse to argumentation and proof. This is why its incomprehension of the problems of scientific discourse is accompanied by a certain tolerance: it approaches such discourse primarily as a variant in the family of narrative cultures. The opposite is not true. The scientist questions the validity of narrative statements and concludes that they are never subject to argumentation or proof. He classifies then as belonging to a different mentality: savage, primitive, under-developed, backward, alienated, composed of opinions, customs, authority, prejudice, ignorance, ideology. Narratives are fables, myths, legends, fit only for women and children. At best, attempts are made to throw some rays of light into this obscurantism, to civilize, educate, develop... (1987: 80).

[5] 'Manufacture' is used in the sense of anything altered by humans from its original form or culturally defined original form. As such it can be applied to components of structure, infrastructure and superstructure of a society in the context of the discussion here.
[6] Pots formed by moulding paste to the interior or exterior of baskets.

into the traditional definition of taphonomy in that the loss of information result from forces or filters in nature that are mechanical, chemical, or biological processes of destruction.

Elements of information from the cultural domain, such as the selection of a specific animal for a specific bone for the manufacture of a specific type of point, are from the realm of 'knowing'. In the 'knowing', these elements belong to and use sets of assumptions that cause the maker of the object to be recognised as a member of their own group. They are part of a cultural tradition that has both the synchronic and the diachronic embed in it. Such objects would have been recognised as of domestic origin and may have had unique indicators that marked them as the product of a specific person. For example Hill (1978: 245) stated that "individual motor-performance variability in the manufacture of artifacts may permit archaeologists to discover which artifacts were made by which specific prehistoric individuals". This is not outside the realm of possibility when we consider that even with the standardization of manufacturing processes after the Industrial Revolution items can retain unique qualities that tie them to specific manufacturers, factories and even machine runs in these factories.

In the making of things, now and in the past, choices for the execution of a particular action fit purposes defined in terms of categories of task, or gender, or age, or categories of the sacred versus the profane or in terms yet unimagined by us. These choices are part of the meaning of the object just as much as the facts of object's shape, size, colour, or any other characteristics that can be quantified, and frequently are quantified, by archaeologists in search of meaning.

The ideational system is subject to the loss of information just as the aspect of concrete-ness may be diminished by the filters of the natural taphonomic process. However, here the forces or filters are cultural. The cultural instructions on how things were done, who did them, or a myriad of other 'facts' are subject to the loss of individual and cultural memory. This is the Doppler Effect as presented by Dethlefsen and Deetz (1965). Of all the associated fact and fiction, truth and fancy embedded or surrounding an 'object', without the instruction manual we end up with the artefact - all hardware, seldom if ever software and absolutely no wetware. We are then in the realm of narrative archaeology (Joyce 2002). This is the breakdown in the transmission of cultural information through the generations. It can be complete or partial and, using the analogy of a biological mutation, it can be just as effective in implementing change in cultural and societal systems as biological mutation is in effecting change in species over time. One of the problems is to pin-point the place at which the information code changed or was lost completely, much like looking for the missing link in the evolution of humans.

The information embedded in the object itself is encoded. Some of the code is decipherable by neither present members of the culture nor members of the academy since the encoding is much like the children's game of telephone: in one end of the 'telephone' line goes the message along to the ear of another who in turn passes it down the line; the resulting message is NEVER exactly the same and if the 'line' breaks there will be no message emerging at the other end. The broken line can also afford for the development of 'just-so' stories on what things meant in the past. These stories can become embedded meaning on the culture of the past by the culture of the day. They can emerge as fantastical mythic constructs that contain information on the development of the metaphorical in a specific culture but actually imparting nothing on the actual past of the material object itself. This is not to say that such contrived meanings do not have their place, principally in the milieu of disrupted cultures attempting to survive the colonial experience.

Ultimately, cultural filters will determine how the researcher will discriminate in the retrieval process and, in analysis, how the recovered materials will be interpreted. Now this is not the final analysis as the realms of narrative archaeology and agency attest. Here are the skills of the researcher informed by the current concepts of method, theory, and ideology within the academy itself. Here too is the startling fact that this can be the level of most information loss, with improper field techniques being the instance of greatest immediate danger – but then would we even notice what we missed? Not likely or at least not until new methods show us the 'missing'! Properly gathered, catalogued, and stored materials can always be subjected to subsequent re-analysis (how scientific!), but re-excavation of the same field unit is impossible. The excavation unit is unique, truly a one-off. As well the explicit, detailed reporting of field techniques, application of these techniques, and the subsequent successes and problems within the fieldwork itself would be of assistance to the eventual evaluative interpretation of the materials. This is most particularly needed in the case of biological materials since they are not chemically inert like stone and properly fired pottery. The counter argument is the lack of funds and time to do reporting of this nature; but then all science must be explicit or its validity comes into question. The foregoing criticism fully recognises that field techniques and concepts of what comprise data change through the years. Nonetheless, there is a 'taphonomy' of information and information systems brought about by the move from the natural to the manipulated, synthetic level. Like the taphonomy of the concrete it has a historical component, bound as it is in time and circumstances.

Final Comments

Traditional studies of the taphonomic process (Behrensmyer and Hill 1980; Lyman 1985; Andrews 1990) examine natural filters. Cause and effect discussions explain, somewhat, the "severe losses of biological information [...] during the taphonomic interval" (Olsen 1980: 5). However, for the archaeologist, unlike the palaeontologist, data that encompass the biological and physical aspects of the world are only part of the retrievable, indeed desired, information. This view diminishes neither the importance, nor indeed the absolute need, for data on the biological and physical environment, nor does it discount their crucial contribution to an understanding of human ecology and thus cultural systems. However, an important issue for the archaeologist is the use of these data in an epiphenomenal way when they may be both inadequate and/or incomplete. Use of such data as though they represent a whole cloth, rather than a patch in an elaborate quilt that remains in construction, contributes to neither an accurate depiction of past life-ways nor an incentive to push beyond the epistemological boundaries delineated by contemporary methodological practices. Additionally, this is exacerbated by the borrowing of theory, methodology, and data from other disciplines, in particular science based ecology, without absorbing their self-critical practice.

Additionally, cultural filters, through which all biological and non-biological concrete items used by humans move, leach away significant information on the original attributes of the concrete itself and its place in any culture under investigation. These cultural filters will discriminate what initially will be selected for use from the environment. They filter the use system itself from concepts of processing through to concepts of discarding 'spent' items. The partial or complete loss of this information affects the outcome of the 'story' but yet may create a dynamic outline for narrative archaeology. However, as archaeologists we should never veer into the unsubstantiated fantastical using inappropriate and ill-informed narrative. Best practice is to be aware of the potential for our own mythic creating process resulting in contrived version of the past presented as definitive interpretation!

The taphonomy of the ideational occurs beyond the boundary of the existing, traditional definition of taphonomy. **Ideational taphonomy** is, thus, an expansion of the concept of taphonomy of the biological to the metaphorical level. It thus imparts the process and its outcome - the loss of information.

Bibliography

ANDREWS, P.
1990 *Owls, Caves and Fossils.* London: Natural History Museum Publications.

BALTENSWEILER, W. and FISCHLIN, A.
1987 On methods of analyzing ecosystems: lessons from the analysis of forest insect systems. In *Ecological Studies, Vol. 61, Analysis and Synthesis Potentials and Limitations of Ecosystem Analysis*, edited by E.D. Schultz and H. Zwolfer. Berlin: Springer-Verlag, pp. 401-415.

BARBOUR, M. G. and BILLINGS, W. D. (eds.)
1988 *North American Terrestrial Vegetation.* Cambridge: Cambridge University Press.

BAKER, B.
2009 For local 'freegans,' dumpsters yield bountiful harvest.http://www.boston.com/news/local/breaking_news/2009/12/for_local_freeg.html [accessed 03.01.2014]

BARNES, R.
1976 Energy. In Hegsted, D. M. (ed.) *Present Knowledge of Nutrition.* NewYork: New York Nutrition Foundation, pp. 10-16.

BARTLEY, W. W., III.
1987 Objective Knowledge and Evolutionary Approach. In Radnitzky, G. and Bartley, W. W. III. (eds.) *Evolutionary Epistemology, Theory of Rationality and the Sociology of Knowledge*, La Salle, Illinois: Open Court, pp. 20-35.

BEHRENSMEYER, A. K.
1978 Taphonomic and ecologic information from bone weathering. *Paleobiology* 4(2): 150-162.

BEHRENSMEYER, A. K., and HILL, A.P. (eds.)
1980 *Fossils in The Making.* Chicago: University of Chicago Press.

BIELAWSKI, E.
1989 Dual perceptions of the past: archaeology and Inuit culture. In Layton, R. (ed.) *Conflict In The Archaeology of Living Traditions*, London: Unwin Hyman, pp. 228-236.

BINFORD, L. R., and BERTRAM, J. B.
1977 Bone frequencies and attritional processes. In Binford, L.R. (ed.) *For Theory Building in Archaeology*, New York: Academic Press, pp. 77-153.

BISHOP, C. A.
1970 The emergence of hunting territories among the Northern Ojibwa. *Ethnology* XI(1): 1-15.

BONNICHSEN, R., and WILL, R. T.
1980 Cultural Modification of Bone: The Experimental Approach in Faunal Analysis. In Gilbert, B. M. *Mammalian Osteology* (revised from *Mammalian Osteo-Archaeology: North America*) Laramie, Wyoming: Modern Printing Company, pp. 7-31.

CASTEEL, R. W.
> 1971 Differential bone destruction: some comments. *American Antiquity* Vol.36(4): 466-469.

CHANEY, M. S. and ROSS M. L.
> 1971 *Nutrition* (8th edition). Houghton Mifflin Company, Boston.

CLARKE, D. L.
> 1968 *Analytical Archaeology*. London: Methuen.

COLLINS, S. L. and WALLACE, L. L.
> 1990 The historic role of fire in the North American grassland. In Collins, S. L. and Wallace, L. L. (eds.) *Fire in North American Tallgrass Prairies*, Norman: University Of Oklahoma Press, pp. 8-18.

COOPER, J. M.
> 1939 Is The Algonquian Family Hunting System Pre-Columbian? *American Anthropologist* 41(1): 66-90.

DAVID, B. and LOURANDOS, H.
> 1998 Rock Art and Socio-Demography in Northeastern Australian Prehistory. *World Archaeology* 30(2): 193- 219.

DAWSON, K. C. A.
> 1975 The Western Area Algonkians. In Proceedings of The Algonkian Conference. Canadian Ethnology Service, *National Museum of Man Mercury Series*, No. 23. Ottawa: National Museums of Canada, pp. 30-41.
> 1976 Algonkians of Lake Nipigon: An Archaeological Survey. *National Museum of Man Mercury Series*, No. 48. Ottawa: National Museums of Canada.

DELANY, M. J.
> 1982 *Mammal Ecology*. Glasgow: Blackie.

DELCOURT, P. A. and DELCOURT, H. R.
> 1981 Vegetation maps for eastern North America: 40,000 Yr. B.P. to the present. In Romans, R. C. (ed.) *Geobotany II* New York: Plenum Press.
> 1987 Long-term forest dynamics of the temperate zone. *Ecological Studies: Analysis and Synthesis*, Vol. 63. New York: Springer-Verlag.

DENNELL, R. W.
> 1979 Prehistoric diet and nutrition: some food for thought. *World Archaeology* 11(2): 121-135.

DENSMORE, F.
> 1929 Chippewa Customs. *Bureau of American Etyhnology* Bulletin 86. Washington: Smithsonian Institution.

DETHLEFSEN, E. and DEETZ, J.
> 1965 The Doppler Effect and Archaeology: A Consideration of the Spatial Aspects of Seriation. *Southwestern Journal of Anthropology* 21(1): 179-275.

DOBRES, M-A. and ROBB, J. E.
> 2002 *Agency in Archaeology*. London: Routledge.

Dods, R. R.
> 1979 *Allotment of Protein by Status: Some African Insights.* Paper presented at the Canadian Archaeological Association Annual Meeting, Spring 1979, Vancouver, British Columbia, Canada.
> 2000 Boundary Markers, Cultural Divisions, and Economic Landscapes. In Malaher, D. G. (ed.) *Selected Papers of Rupert's Land Colloquium 2000*, Winnipeg Centre for Rupert's Land Studies at the University of Winnipeg. Pp. 69-83.
> 2002 The Death of Smokey Bear: the ecodisaster myth and forest management practices in prehistoric North America. *World Archaeology* 33(3): 475-487.
> 2003 Wondering The Wetland: archaeology through the lens of myth and metaphor in Northern Boreal Canada. *Journal of Wetland Archaeology* 3: 17-36. Oxbow Books.
> 2004 Knowing Ways/Ways of Knowing: Reconciling Science and Tradition. *World Archaeology* 36(4): 547-557.
> 2007a Intersubjectivity and the Meaning of Things. *The International Journal of the Humanities* (Common Ground). http://ijh.cgpublisher.com/product/pub.26/prod.933 [acessed 02.01.2014]
> 2007b Pyrotechnology and Landscapes of Plenty in The Northern Boreal. Chapter 9. In Gheorghiu, D. and Nash, G. (eds.) *The Archaeology of Fire: Understanding Fire as Material Culture*. Budapest: Archaeolingua.
> 2009 The Syntax of Place and Space. Chapter 4 In Nash, G. and Gheorghiu, D. (eds.) *Archaeology of Territoriality and People*. Budapest: Archaeolingua.

DOWNING, J. A.
> 1991 Comparing apples with oranges: methods of interecosystem comparison. In Cole, J. J., Lovett, G. M. and Findlay S. E. G. (eds.)*Comparative Analysis of Ecosystems: patterns, mechanisms, and theories*. New York: Springer-Verlag, pp. 24-45.

DUDEN, B.
> 1991 *The Woman Beneath The Skin: A Doctor's Patients in Eighteenth-Century Germany*. Cambridge, Mass.: Harvard University Press.

DUNNELL, R.C.
> 1989 Philosophy of science and archaeology. In Pinsky, V. and Wylie, A. *Critical Traditions in Contemporary Archaeology: essays in the philosophy, history and socio-politics of archaeology*. Cambridge: University Press, pp. 5-9.

EMBREE, L.
> 1989 The structure of American theoretical archaeology: a preliminary report. In Pinsky, V. and Wylie, A. (eds.) *Critical Traditions in Contemporary Archaeology: essays in the philosophy, history and socio-politics of archaeology*. Cambridge: University Press, pp. 28-37.

FARKAS, C. S.
> 1979 *Survey of Northern Canadian Indian Dietary Patterns and Food Intake*. Man Environment Studies. Waterloo: University of Waterloo.

FEYERABEND, P.
> 1988 *Against Method*. New York: Verso. (revised edition)
> 1994 Realism. In Gould, C. C. and Cohen, R. S. (eds.) *Boston Studies in the Philosophy of Science*, Vol. 154. Dordrecht: Kluwer Academic Publishers, pp. 205-222.

GIFFORD, D. P.
 1978 Ethnoarchaeological observations of natural processes affecting cultural materials. In Gould R. A. (ed.) *Explorations in Ethnoarchaeology*. Albuquerque: University of New Mexico Press, pp. 77-101.
 1980 Ethnoarchaeological contributions to the taphonomy of human sites. In Behrensmeyer, A. K. and Hill, A. P. *Fossils in the Making*. Chicago: University of Chicago Press, pp. 93-106.

GOULD, C. S. and COHEN, R. S. (eds.)
 1994 Artifacts, Representations and Social Practice. *Boston Studies in the Philosophy of Science*, Vol. 154. Dordrecht: Kluwer Academic Publishers.

HANEN, M .P., OSLER, M. J., and WEYANT, R. G. (eds.)
 1980 *Science, Pseudo-Science and Society*. Calgary Institute for the Humanities. Waterloo, Ontario: Wilfred Laurier University Press.

HAWKING, S. W.
 1988 *A Brief History of Time*. Toronto: Bantam Books.

HEAL, O. W. and GRIME, J. P.
 1991 Comparative analyses of ecosystems: past lessons and future directions. In Cole, J., Lovett, G., and Findlay, S. (eds.) *Comparative Analysis of Ecosystems: Patterns, Mechanisms, and Theories*. New York: Springer-Verlag, pp. 7-23.

HEGSTED, D. M. (ed.)
 1976 *Present Knowledge of Nutrition*. New York: New York Nutrition Foundation.

HICKERSON, H.
 1967 Land Tenure of The Rainy Lake Chippewa At The Beginning of The 19th Century. *Smithsonian Contributions to Anthropology*, 2(4). Smithsonian Institution, Washington.

HILL, A.
 1976 On carnivore and weather damage to bone. *Current Anthropology*, 17(2): 335-336.

HILL, J. N.
 1978 Individuals and Their Artifacts: An Experimental Study in Archaeology. *American Antiquity*, 43(2): 245-257. Contributions to Archaeological Method and Theory.

HILLS, G. A.
 1959a Soil-forest relationship in the site regions of Ontario. *First North American Forest Soils Conference*. Bulletin. East Lansing: The Agricultural Station.
 1959b *A Ready Reference to The Description of The Land of Ontario and Its Productivity*. Preliminary Report. Maple: Ontario Department of Lands and Forests.
 1976 An integrated iterative holistic approach to ecosystem classification. *Proceedings, Meeting of the Canadian Commission on Ecological Land Class*, Petawawa.

HØYGAARD, A.
 1941 *Studies on The Nutrition and Physio-Pathology of Eskimos*. Skrifter utgitt au Det Norske. Videnskaps-Akademi I Oslo. I. Mat. Naturv. Klasse. 1940, No. 9. Oslo: I Kommisjon Hos Jacob Dybwad.

HUDSON, S.
 2009 Dumpster diving with a culinary purpose. www.westword.com/restaurants/dumpster-diving-with-a-culinaary-purpose-5728131 [accessed 22.01.2017]

HURLEY, W. M.
 1979 *Prehistoric Cordage: Identification of Impressions on Pottery*. Aldine Manuals On Archaeology - 3. Washington: Traxacum.

ILLICH, I.
 1971a *Celebration of Awareness*. London: Calder & Boyas.
 1971b *Deschooling Society*. http://ecotopia.com/webpress/deschooling.htm [accessed 22.01.2017]
 1978 *Toward a History of Need*. New York: Pantheon Books.
 1982 *Gender*. New York: Pantheon Books.
 1988 ABC: The Alphabetization of the Popular Mind. San Francisco: North Point Press.
 1992a *In the Mirror of the Past*. London: Marion Boyars Publishers.
 1992b *Ivan Illich in Conversation* interviews with Cayley, David. Toronto: Anansi Press.

JOYCE, R.
 2002 *The Languages of Archaeology: Dialogue, Narrative, and Writing*. Hoboken, NJ: Wiley-Blackwell.

KREBS, C. J.
 1985 *Ecology: The Experimental Analysis of Distribution and Abundance*. (3rd edition). New York: Harper Publishers.

LAFORD, R.
 1958 Some soils, vegetation and site relationships to the climatic and sub-climatic black spruce forests in Northeastern America. *First North American Forest Soils Conference*. Bulletin of The Agriculture Station, Michigan. East Lansing: The Agriculture Station, pp. 670-674.

LANGE, F.W. and RYDBERG, C. R.
 1972 Abandonment and post-abandonment behaviour at a rural Central American house site. *American Antiquity*, 37(3): 419-432.

LÉVI-STRAUSS, C.
 1964 *Le cru et le cuit*. Paris: Libraire Plon.

LOESCHCKE, V.
 1987 Niche structure and evolution in ecosystems. In Schultz, E. D. and Zwolfer, H. (eds.) *Ecological Studies, Vol. 61, Analysis and Synthesis Potentials and Limitations of Ecosystem Analysis*. Berlin: Springer-Verlag, pp. 320-332.

LUGENBEAL, E.
 1978 The Blackduck ceramics of the Smith site (21KC3) and their implications for the history of Blackduck ceramics and culture in northern Minnesota. *Mid-Continental Journal of Archaeology*, 3(1): 42-67.

LYMAN, R. L.
 1979 Available meat from faunal remains: a consideration of techniques. *American Antiquity* 44(3): 536-546.

1985 Bone frequencies: differential transport, *in situ* destruction, and the MGUI. *Journal of Archaeological Science* 12: 2221-236.

1994 *Vertebrate Taphonomy*. Cambridge: Cambridge University Press.

LYON, P. J.
1970 Differential bone destruction: an ethnographic example. *American Antiquity* 35(2): 213-215.

LYOTARD, J.-F.
1987 The Postmodern Condition. In Baynes, K., Bohman, J., and McCarthy, T. (eds.) *After Philosophy: End or Transformation?* Cambridge, Massachusetts: The MIT Press.

MacDONALD, G. M.
1984 *Postglacial plant migration and vegetation development in the western Canadian boreal forest*. Dissertation, University of Toronto. Toronto.

1987a Postglacial vegetation history of the Mackenzie River Basin. *Quaternary Research* 28: 245-262.

1987b Postglacial development of the subalpine-boreal transition forest of western Canada. *Journal of Ecology*, 75:303-320.

McLEOD, M.
1978 *The Archaeology of Dog Lake, Thunder Bay: 9000 Years of Prehistory*. Report MS submitted to The Ontario Heritage Foundation. Toronto.

MORLAN, R.
1980 Taphonomy and archaeology in the upper Pleistocene of the northern Yukon Territory, a glimpse of the peopling of the New World. *National Museum of Man Mercury Series,* No. 94. Ottawa: National Museum of Man.

OLSEN, E. C.
1980 Taphonomy: its history and role in community evolution. In Behrensmeyer, A. K. et al. (eds.) *Fossils In The Making*, edited by Chicago: University of Chicago Press, pp. 5-19.

OSLER, M. J.
1980 Apocryphal knowledge: the misuse of science. In Hanen, M. P., Osler, M. J., and Weyant, R. G. (eds.) *Science, Pseudo-Science and Society*. Calgary Institute for the Humanities. Waterloo, Ontario: Wilfred Laurier University Press, pp. 273-290.

PAYNE, S.
1972a Partial recovery and sample bias - the results of some sieving experiments. In Higgs, E. S. (ed.) *Papers in Economic Prehistory*. Cambridge: Cambridge University Press, pp. 49-64.

1972b On the interpretation of bone samples from archaeological sites. In Higgs, E. S. (ed.) *Papers in Economic Prehistory*. Cambridge: Cambridge University Press, pp. 65-81.

PETERS, R. H.
1991 *A Critique For Ecology*. Cambridge: Cambridge University Press.

PIEPENBRINK, H.
1986 Two examples of biogenous dead bone decomposition and their consequences for taphonomic interpretation. *Journal of Archaeological Science*, 13: 417-430.

POERKSEN, U.
1995 (1988) *Plastic Words:* The Tyranny of a Modular Language. University Park, PA: Pennsylvania State Press.

POLANYI, M.
1966 *The Tacit Dimension*. New York: Doubleday & Co.

POPPER, K. R.
1978 Three Worlds. *The Tanner Lecture on Human Values*. The University of Michigan. April 7, 1978. http://tannerlectures.utah.edu/_documents/a-to-z/p/popper80.pdf [accessed 22.01.2017]

1983 *Realism and The Aim of Science*. Totowa, New Jersey: Rowman and Littlefield.

POTZGER, J. E.
1946 Primeval forests in central northern Wisconsin and upper Michigan and a brief post-glacial history of the Lake Forest Formation. *Ecological Monographs*, 16: 211-250. Durham: Duke University Press.

PRICE., T. D.
1985 The reconstruction of Mesolithic diets. In Bonsall, C. (ed.) *The Mesolithic in Europe*. Edinburgh: John Donald Publishers, pp. 48-59.

REICHLE, D. R. (ed.)
1973 *Analysis of Temperate Forest Ecosystems*. Berlin: Springer-Verlag.

ROGERS, E. S. and BLACK, M. B.
1976 Subsistence strategy in the Fish and Hare Period, Northern Ontario: the Weagomow Ojibwa, 1880-1920. *Journal of Anthropological Research* (SWJA), 32(1): 1-43.

Saunders, J. J. and Daeschler, E. B.
1994 Descriptive Analyses and Taphonomical Observations of Culturally-Modified Mammoths Excavated at "The Gravel Pit," near Clovis, New Mexico in 1936 . *Proceedings of the Academy of Natural Sciences of Philadelphia*, Vol. 145: 1-28 .

SCHIFFER, M. B.
1972 Archaeological context and systemic context. *American Antiquity*, 37(2): 156-165.

1987 *Formation Processes of the Archaeological Record*. Albuquerque: University of New Mexico Press.

SCHULTZ, E. D. and ZWOLFER, H. (eds.)
1987 Analysis and Synthesis Potentials and Limitations of Ecosystem Analysis. *Ecological Studies*, Vol. 61. Berlin: Springer-Verlag.

SHUGART, H. H., LEEMANS, R., and BONAN, G. B. (eds.)
1992 *A Systems Analysis of the Global Boreal Forest*. Cambridge: Cambridge University Press.

SMITH, B. D.
1976 'Twitching': a minor ailment affecting human paleoecology research. In Cleland, C. A. *Culture Change and Continuity*. New York: Academic Press, pp. 275-292.

SPETH, J. D.
 1983 Bison Kills and Bone Counts. *Prehistoric Archaeology and Ecology Series*. Chicago: University of Chicago Press.

TAMM, C. O. (ed.)
 1976 Man and The Boreal Forest. *Proceedings of a regional meeting within Project 2 of MAB (UNESCO's Man and the Biosphere Programme)*. Stockholm: Ecological Bulletin (Stockholm) 21.

TURNER, M. G., and GARDNER, R. H.
 1991a Quantitative methods in landscape ecology: an introduction. In Turner, M. G and Gardner, R. H. (eds.) *Quantitative Methods in Landscape Ecology: the analysis and interpretation of landscape heterogeneity*. New York: Springer-Verlag, pp. 3-14.
 1991b (eds.) *Quantitative Methods in Landscape Ecology: the analysis and interpretation of landscape heterogeneity*. New York: Springer-Verlag.

VENNUM, T., Jr.
 1988 *Wild Rice and The Ojibway People*. St. Paul: Minnnesota Historical Society Press.

WHEELER, J. A.
 1990 *A Journey into Gravity and Spacetime*. New York: Scientific American Library.

WHEELER, Sir M.
 1954 *Archaeology From The Earth*. London: Oxford At The Clarendon Press.

WILLIAMS, J. T.
 1992 Life tables in palaeodemography: a methodological note. *International Journal of Osteoarchaeology*, Vol. 2: 131-138.

WING, E. S. and BROWN, A. B.
 1979 *Paleonutrition*. New York: Academic Press.

WRIGHT, J. V.
 1967a The Laurel Tradition and Middle Woodland Period. *National Museums of Canada Bulletin*, No. 217. Ottawa: National Museums of Canada.
 1968b The Michipicoten Site, Ontario. Contributions to Anthropology VI. *National Museums of Canada Bulletin*, No. 224, Ottawa: National Museums of Canada, pp. 1-85.

Sustainability, Health, and Society: Prehistoric Artefacts as Sustainable Materials

Lolita Nikolova

Abstract

This research favours multifunctional differentiation of the archaeological fragmented material (waste, recycling, secondary use, ritual, etc.). It proposes that the fragmented pottery served a complex function as a sustainable material in prehistoric settlements. The pottery perhaps reveals different aspects of village activity and concepts about the ancient natural and social environment. It remains unclear what archaic people's awareness was about processes that today we describe as recycling and secondary use, and it remains unclear whether reusing pottery made a difference in their lives. The prehistoric villages were places for production and reproduction of material culture, some components of which could have been visibly changed and re-incorporated with a different function. This kind of activity possibly integrates both modern characteristics of recycling and secondary use.

This work also points out the importance of developing an archaeology of health as a branch that will help not only to learn more about humankind's health strategies from the past, but also to strengthen and develop methodologies of archaeological excavations now that will extract more anthropological information from the excavated layers. It is a step in the anthropologization of interpretive archaeology through theoretical modelling and interpretation of the archaeological evidence as records about people's everyday life from the perspectives of sustainability and health.

Introduction

Prehistory is the longest period in human history marked by the absence of literary records. One of the reasons this period was so long was likely due to the fact that people were engaged in successful reproduction strategies that obviously satisfied their daily lives and encouraged living as communities by regulating differences and general changes based on a strong, verbal, social memory. Another way to think about it is to believe that the power of traditions was incorporated into the roles and the actual power of their leaders.

Thus, studying in depth the everydayness (i.e., daily routines) of prehistoric people helps to not only better learn their material culture, but also their way of thinking and their motivations. As our contemporary world shows, tradition, in many forms, is embodied in people's everydayness and usually innovations in traditions impact only specific aspects of the social life. Briefly speaking, all human societies are traditional societies from the perspective of the majority or the popular culture. We even can think of culture as being born popular.

Reusing pottery is one of the most popular prehistoric social practices. There are many ways in which this activity has been documented. Theoretically, the problem can be defined as whether there is a difference between recycling and secondary use in prehistory (e.g. Schiffer 1987; Ellis 2000). The variable definitions of recycling include either broad modification and reuse or remanufacturing and reuse of artefacts, the secondary function of which is different than the original one. Secondary use means a change in the function, but not in the morphology of the objects (Ellis 2000: 43). None of these frameworks include the cases, in which we do have reuse of pottery with the same function. Such reuse is best documented by holes in the walls of the vessels, which were obviously broken and connected with a cord (repairing status). Then, there are three aspects of activity (beyond production and wasting) attributed to the sustainable society: repairing, secondary use, and recycling. Since every society has components of sustainability, the degree of sustainability depends on the character of the five fields of activities and the interrelation between them:

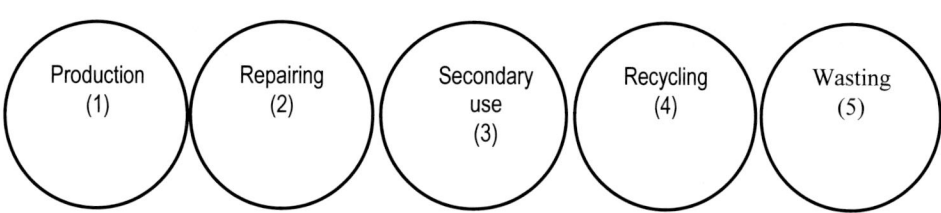

Scheme 1: *Societal components related to sustainability: production (1), repairing (2), secondary use (3), recycling (4) and wasting (5).*

Using empirical data from the prehistoric Balkans, we will focus on three strains of secondary/recycling use of pottery in prehistory of the Balkans and will try to specify the role of the fragmented pottery in the everyday life of the prehistoric communities.

Fragmented Pottery in Villages

Balkan prehistoric sites provide some of the richest records for world prehistory (Bailey 1999, 2000, 2005; Nikolova 1996, 1999). Unfortunately, the numerous examples of fragmented pottery have been neglected even in the publication of whole sites and there are missing contextual and functional publications for in-depth research. Hopefully, this research will stimulate new publications that will clarify basic questions, in particular, the specific role of the different kinds of pottery as sustainable material.

It can be proposed that there are three main strains of fragmented pottery in the prehistoric sites:

Strain 1. Fragmented pottery without a specific building function.
Strain 2. Ceramic spots and stripes that may indicate paths in the villages.
Strain 3. Fragmented pottery with a specific building function (e.g. placed on the floors of features like hearths and ovens, possible fills of postholes, etc.).

Strain 1

This sort has the least-defined functional characteristics and includes the fragmented pottery discovered in the cultural layers that does not have a specific residential or building function. When there is a vertical stratigraphy, usually the pottery in the layer between two building levels has been attributed to the lower layer. But, the fact that, for instance, in the Early Bronze layer, there can be found even Later Copper Age fragmented pottery shows that the character of the pottery is complex and cannot be connected directly with the stratigraphy. The most popular case is with Rachmani pottery at Pevkakia Mogula, which was ascribed to the Early Helladic period because the complex character of the prehistoric people's activity on the site was not taken into account (Weisshaar 1996).

In Bulgaria, there are cases of Later Copper Age fragmented pottery in the Early Bronze layers (Dubene-Sarovka), as well as instances of Neolithic pottery in Early Bronze Age layers (Vesselinovo) (Nikolova 1999: 32 and references cited there).

There are many ways to explain the earlier pottery in later layers. It could come by:

A. Digging pits in the village
B. Digging ditches around or in the village

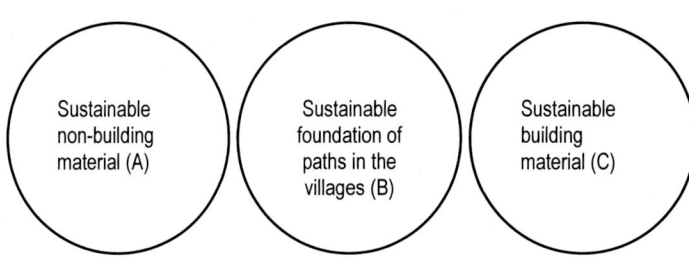

Scheme 2: *Prehistoric fragmented pottery as a sustainable material: non-building (A), foundation of paths (B), and building material (C).*

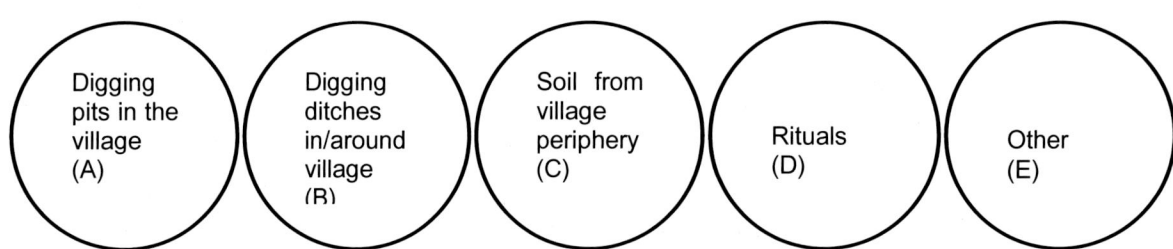

Scheme 3: *Explanatory models of obtaining of earlier prehistoric fragmented pottery for use in later levels: by digging pits in the village (A), digging ditches in or around the village (B), obtaining soil from the periphery of the village (C), rituals (D) or other (E).*

C. Taking soil from the periphery of the village to establish a new level of the new village or for partial renovation.

D. Rituals related to social memory, continuity, and genealogy, etc.

In the case of Dubene-Sarovka, a Late Copper Age village was documented in the southern area of the site (Nikolova 1999: 28). Then, there is no reason not to connect the Later Copper Age potsherds discovered in the Early Bronze Age levels with that village. The potsherd number is not considerable, so the best explanation would likely be to associate them with taking soil from the periphery of the village for establishing a new level and partial renovation (Model C). In addition, the fact that the quantity of the later Copper Age potsherds is minimal may show a purposeful, specific (symbolic?) use (Model D).

It has been posited in the specialized literature for different archaeological periods that the pottery was in secondary use, because of the opportunity to absorb water. Rain was an extremely large problem for prehistoric villagers. Because of the wattle-and-daub house constructions in the Balkans, the rain and damp damaged the houses by stimulating mould and wood decay. We have many cases of partial renovations of the houses and one of the reasons could be the damaging effect of rain. Successful social strategies were sought for dealing with too much rain and one of them was by inclination of the village from north to south on tells, for instance at the Yunatsite tell (Катинчаров, Мерперт, Титов, Мацанова & Авилова, 1995: Annex 1). Some of the houses were also fired and rebuilt in the same place (Gheorghiu, 2016, with the references cited there). Unfortunately, such social activities are documented in ways that allow for controversial interpretations.

Strain 2

This sort would include ceramic spots and stripes that may indicate paths in the villages. Interestingly, people mostly used pottery on the tells, whilst stone paving was an exception - see e.g. Dyadovo (Sekime and Kamuro 2000: Pl. 5). This fact can be explained by the character of the cultural layer, which was in some cases amorphous, as well as with the drainage function of pottery.

Strain 3

Fragmented pottery was used as a sustainable material with specific building functions. The typical prehistoric mode is fragmented pottery embedded in the floors of features such as hearths and ovens. The pottery could be derived from one or more vessels (in some cases luxury/richly ornamented) or just secondary used potsherds from different vessels. It is possible to think about a sort of secondary symbolization by placing a richly ornamented recently broken vessel in the floor of the hearth as a sacrifice to ancestors or of the hearth goddess. The mechanism of building of hearths in the prehistoric houses is still not very well understood. Since there are cases when several hearths were built in one house but functioned at different periods, it is possible to interpret them as evidence of the ritualization of domestic space, such as: destroying a hearth after a death of household member, building a new hearth after the birth of a new member of the household or after a marriage and welcoming the new couple into the home of parents, or a change in the occupants of a house. Unfortunately, the static model dominates in the village prehistoric anthropology, and circulation of people is discussed mostly in migration studies, although village life had its own dynamics that are visibly or invisibly documented in the archaeological record.

Numerous instances of ceramic floors of ovens have been published from Dyadovo (e.g. Sekime and Kamuro 1998, 1999, 2000; Kamuro 2006). As the excavations at Dubene-Sarovka showed, similar ceramic floors were typical of both the Yunatsite and Ezero cultures during the Early Bronze Age.

It is also possible that the soil was mixed with fragmented pottery by the filling of postholes of some prehistoric buildings.

A complex number of reasons can be proposed to explain the richness of fragmented pottery in the cultural layers:

(1) Breaking pottery during circulation is possibly one of the main reasons. Pottery itself is not a very steady material, since it is fragile.

(2) Recycling pottery because of hygiene reasons. Prehistoric pottery contains different porosity that becomes a bacteria source.

(3) Fashion. The prehistoric pottery belongs to different style and the adoption of a new style can be one reason for recycling the old-fashioned pottery.

(4) Breaking vessels during feasting. This pattern is well known in Greek culture.

(5) When members of the household left the house. There are documented cases of people treasuring the pottery in destroyed houses. In other cases, the pottery would just be recycled if not taken to the new home.

(6) Using pottery for special rituals and recycling it after the ritual.

(7) Overusing the pottery and recycling because of burning, scratches, wear, etc.

(8) Defective production. There are many documented cases of defective pottery production in prehistory.

(9) Purposeful filling the layer with fragmented pottery for drainage, for instance.

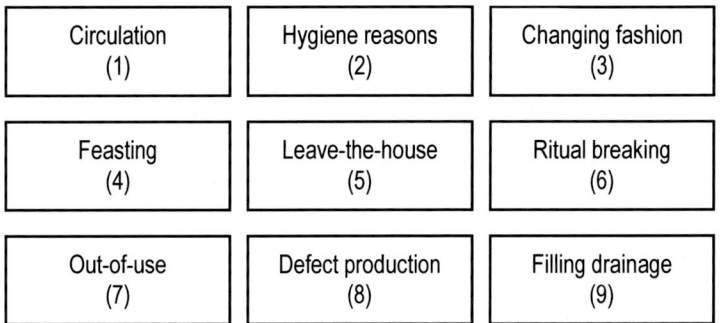

Scheme 4: *Modelling of possible reasons for finding fragmented pottery in prehistoric levels*

Sustainable Materials and Prehistory

The problem of sustainable materials in prehistory can be approached in similar way to the problem of sustainability in the contemporary world – about the [balanced] interrelation between environmental protection and economic development (Blackburn 2007: 2) and an opportunity for production and reproduction of a healthy life.

Tell style of life and sustainability

There is no reason not to believe that this problem of sustainability existed from the beginning of the human civilization, since people's interaction with nature is always ambiguous and ambivalent.

One of the solutions in prehistory was to build consecutively layered settlements (tells). Most of them are known based on partial excavations while the entirely excavated tells, like Golyamo Delchevo and Ovcharovo, do not give reliable information for a scientific analysis. The artificial character of the tell cake-model is very clear for instance, due to the fact that archaeological material attributed to "different phases" by H. Todorova were illustrated in one and the same table. The absence of published depths of the finds and how they relate to different houses shows that in the past it was possible that the academic literature was used for non-scientific and misleading publications. The most important requirement for any scientific findings is verification.

Tells, such as Ezero and Yunatsite, based on professional excavation of international teams of scholars, reveal complex stratigraphy with peculiarities. For instance, at Yunatsite, it was documented in the field that eventually the people took purposeful social strategies to control rain water (see above), while at Ezero, excavations showed that only parts of the houses are preserved on the site, which makes it difficult to precisely reconstruct plans and requires multi-variant interpretations of the findings. The same situation was documented at Dubene-Sarovka. The recent Dyadovo excavations also indicate the fragmentary character of preservation in the architectural remains.

Does the emergence of the tell as a way of living relate to the problem of environmental protection and economic development of the early agricultural-stockbreeding communities? To answer this question, we need to presume that for instance: 1. Despite the available huge natural space for agriculture, the early agricultural communities had preferences toward a complex set of preconditions for settlement – soil, water, communication access, wood, cultural traditions, etc. 2. They attempted to reproduce the established, specific networks of relationships that required environmental protection. 3. They recognized that environmental protection supported the economic development that in turn stabilized strategies that included sustainable materials in everyday activities.

The archaeological argument of the first presumption is the very stable settlement networks during the Neolithic, Copper Age, and Early Bronze Age in the Balkans. The changes are usually within existing micro-systems that may reflect local changes like the alteration of the bed of the river. For instance, the distance between the Neolithic and Copper –Early Bronze Age settlement-tells, near the present village of Dubene (Karlovo Municipality, South Bulgaria) is about 5 km and possibly reflect a change in the course of the river Stryama from north to south.

The environmental-economic awareness of the prehistoric communities is probably one of the main causes for the abandonment of some of the tells for a certain period of time or permanently. Recognizing such awareness from the beginning even of the human history of the Balkan population is very essential in understanding later prehistoric processes like the cultural changes in the fourth millennium cal BCE. Environment and economy are fundamental constructs of social identity and can directly and indirectly influence the entire life style.

The second presumption connects the anthropology and ecology of place. As today, in prehistory, living in one place was a problem. There were different potential strategies to resolve this problem. In many places in the Balkans, the resolution was the tell (Southern Bulgaria and Lower Danube basin, North Greece, Macedonia, and European Turkey). Keeping the tell as a comfortable place of living required not only reproducing positive social relationships, but also an environment that would prevent for instance, death and destruction. Fragmented pottery discovered in the inter-building levels was one of the sustainable materials that strengthened the layer by the pottery's ability to absorb water, and at the same time, it was a place of recycling this huge amount of what would otherwise be waste material.

The third presumption (that environmental protection supported economic development) can be argued by recalling the longevity of the tells.

Healthy Life and Sustainability

A history of Balkan prehistoric health is still missing. Health status has been applied mostly to palaeopathology.

The construction of a general concept of health in Balkan prehistory requires the building of a scientific framework for the better understanding our distant ancestors by departing both from non-scientific intuitive and misleading materials published in academic literature and from meaningless theories. The goal, however, is limited by the poor quality of record base.

Beginning with the preconditions for healthy life, the natural environment and climate were excellent factors for a long healthy life in the Balkans. The water resources and the rich soil for agriculture and livestock pastures at different elevations in the Balkan Mountains were preconditions for a successful reproductive subsistence economy based on agriculture and stockbreeding. However, the high death rates of children and adults and the absence of many burials of elderly individuals suggest that life was a challenge and the prehistoric population had to resolve many problems, including a possible high rate of illness. As the contemporary traditional societies show, in such a situation, health becomes a prominent concern.

A variety of archaeological records can be used for understanding the health model. The settlement patterns indicate that it was the possible long-term occupation of one and the same place that resulted in tells (see above). The tells were places where developed communities reproduced a healthy-oriented culture.

It can be presumed that the traditional medicine in Balkan prehistory was based on herbs for treating most of the diseases, since the region is extremely rich in medicinal herbs. The choice of herbs depended on the area's local resources, although the exchange and search for exotic herbs, as well as the gradual increase in knowledge in traditional medicine cannot be excluded.

Ethnographic models help to better understand how the people accepted and used medicinal flora. For instance, in Trinidad, bush medicine was valued as traditional wisdom whose "ready availability in villages like Pinnacle argues for superiority of rural life over that in town". Such an emic view confirmed that traditional cultures are health-oriented and the cultural status of all aspects of preventive health and remedies was high. Every adult in the village of Matelot "has a working knowledge of some bush and most can describe the properties of between thirty and a hundred". About twenty of the plants were in common use. Everyone has favourites, but it is easy to switch from one to another plant in favour of another recommended or supplied by a neighbour. Medicinal herbs are grown in the house yard or easily found in the forest. Causes of sickness are weather and climate (1), conditions of work (2), a change in the hot-cold balance of the body (3), or the neglect of some other health precaution (4). Small quantities of the medicinal herbs used for particular sickness can be taken for its prevention (Littlewood 2007: 20-21).

The stepped "hierarchy of resort" (after Romanucci-Ross 1969) was probably embedded in the prehistoric response to diseases. According to this concept, with developing understanding in the medical anthropological literature, people first try one thing and then try another until satisfied. The patterns of resort "involve many types of treatment, in parallel or pluralistically, at once" (Sobo 2009: 60).

The ethnographic research also points to the prominent role of the women as healers (McClain 1989). Feinerman (1989: 30-32) distinguishes the following health care resources in Saraguro: curanderos, herbalists, midwives, pharmacists, nurses, physicians. The first three categories can be connected to prehistory. Herbalists and midwives can be compared with the Roman "practical" medicine, while the curanderos correspond to the vague level of Etrusco-Latin religious and magical medicine (Scarborough 1969: 25). Further, the Druids interpreted as priest-physicians (Garber 2008: 6), represent the model of healers who were also involved in religious-astronomic practices, as it is evidenced from Pliny's text:

'They believe that anything growing on oak-trees is sent by heaven. The mistletoe was collected on the sixth day of the Moon. Then, greeting the Moon with the phrase that in their own language means healing all things ...' (after Garber 2008: 7)

Then, records from different epochs show that the prehistoric medicine was most probably multilayered and included at least three levels: (1) household level where the women dominated (1), specialized practitioners/witches (2), and priests (3). We also can propose regulation of the food consumption and beverage by norms, rituals, and taboos.

There is not much evidence that can prove this model for Balkan prehistory. Medicinal plants discovered in the household context may prove to be the first category. Candidates for the second and third categories are probably some of the prehistoric female figurines that were multifunctional. One of the possible functions would be magical (healing) and stimulating fertility and healthy life. Some of the figurines could also possibly be interpreted as healers.

There are also dolmens in the Rhodope and Strandhza Mountains, with analogies in the Caucasus (Smith, Badalyan, and Avetisyan, 2009: Plates 78-79). They have been most often interpreted as tombs, although no skeletons were discovered. Distant analogies can be provided with Stonehenge in Europe and the medicine wheels in North America (Garber 2008: 3-6). The data cannot exclude an interpretation that would connect the dolmens and stone circles not only to burial customs but also to healing practices in antiquity. Oak was sacred for Druids, while it is possible the whole Rhodope Mountains were thought of as sacred since they were the birthplace of Orpheus.

Conclusion

Based on the above approach, we can formulate the hypothesis that there are different reasons for finding fragmented pottery in the prehistoric villages, including recycling. It is likely that in prehistory a cognitive understanding of sustainable materials was not developed as it is in the present day (e.g. Blackburn 2007). But our exemplary stains 1-3 are the empirical base to propose that the fragmented pottery was used as a sustainable material and it integrated both the village activity and a concept about the ancient environment. It remains unclear what the awareness was of archaic people about processes that today we describe as recycling and secondary use.

The prehistoric villages were used for production and reproduction of material culture, some components of which could have been visibly changed and re-incorporated with a different function. This kind of recycling possibly integrated the modern characteristics of recycling and secondary use.

Approaching the problem of sustainable materials in prehistory, it is clear that the main body of evidence remains to be documented and published in the future by changing the methodology of documentation and publication of prehistoric sites and stressing attention not only on features and residential artefacts, but also on the sustainable material, the graves fills and on the surrounding environment.

The problem of sustainability is connected to health and this research proposes medical knowledge and health were of primary importance for prehistoric communities, in a similar way as they are to traditional cultures.

Typically, in the past, prehistoric archaeology looked at villages as places where there was accumulated material culture that was studied in detail, but disconnected from the people and environment. There has been a developing process of the anthropologization of archaeology (see e.g. Nikolova 2004a, 2004b; Bailey, Whittle, and Hofmann, 2008), but even diet has been largely researched as an abstract record of subsistence, without a connection with the individuals and households. The advance of research in human society makes it possible to change the agenda through an integrative approach and thus, reconsider every piece of well documented archaeological information, in order to study archaeologically and anthropologically prehistoric sites as places where people exercised and developed social strategies of healthy life, produced and reproduced different types of social relations, and circulation of ideas, as well as practices of enculturation and socialization, and lastly, entertained and developed networks of relationships with close and distance communities. Such an agenda in turn reflects on the methodology of excavation and make it possible to gather much more valuable information from any square centimetre of excavated area by using different techniques and methods from the natural and social sciences.

Bibliography

BAILEY, D. W.
   1999 What is a tell? Spatial, temporal and social parameters. In: Brück, J. and Goodman, M. (eds.), *Making places in the prehistoric world* (pp. 94-111). London: UCL Press.
   2000 *Balkan Prehistory. Exclusion, incorporation and identity*. Routledge. London & New York.
   2005 *Prehistoric figurines: representation and corporeality in the Neolithic*. London and New York: Routledge, Taylor and Francis Group.

BAILEY, D. W., WHITTLE, A., and HOFMANN, D. (eds).
   2008 *Living well together? Settlement and materiality in the Neolithic of south-east and central Europe*. Oxford: Oxbow Books.

BLACKBURN, W. R.
2007 *Sustainability handbook: the complete management guide to achieving social economic and environmental responsibility.* London: Earthscan Publications.

ELLIS, L. (ed.)
2000 *Archaeological Method and Theory: An Encyclopaedia.* New York and London: Garland Publishing Co.

FEINERMAN, R.
1989 The forgotten healers: women as family healers in an Andean Indian community. In C. S. McClain (ed.), *Women as healers. Cross-cultural perspectives* (pp. 24-41). New Brunswick and London: Rutgers University Press.

GARBER, J. J.
2008 *Harmony in healing.* New Brunswick and London: Transaction Publishers.

GHEORGHIU, D.
2016 Building and burning: The construction and combustion of Chalcolithic dwellings in the lower Danube and the eastern Carpathian areas from the perspective of experimental archaeology. In L. Nikolova, M. Merlini, and A. Comşa (eds.), *Western-Pontic Culture Ambience and Pattern. In memory of Eugen Comşa.* Berlin, Boston: Walter de Gruyter GmbH.

KAMURO, H. (ed.)
2006 *Dyadovo excavation 2004. A preliminary report of the 17th excavation at Dyadovo, Bulgaria.* Tokay: Tokai University Thracian Expedition.

LITTLEWOOD, R.
2007 Coconuts and syphilis: an essay in overinterpretation. In R. Littlewood (ed.), *On knowing and not knowing in the anthropology of medicine* (pp. 18-27). Walnut Creek: Left Coast Press.

McCLAIN, C. S. (ed.)
1989 *Women as healers. Cross-cultural perspectives.* New Brunswick and London: Rutgers University Press.

NIKOLOVA, L.
1996 Settlements and ceramics: the experience of Early Bronze Age Bulgaria. In: Nikolova L. (Ed.), *Early Bronze Age settlement patterns in the Balkans (ca. 3500-2000 BC calibrated dates).* Part 2 (pp. 145-186). Sofia: Prehistory Foundation and Agatho. Reports of Prehistoric Research Projects 1.
1999 *The Balkans in later prehistory.* Oxford: BAR. BAR International Series 791.
2004a Notes on the study of early social reproduction in Thrace (Based on data from Neolithic mortuary practices). In: V. Nikolov, K. Bacvarov, and P. Kalchev (eds.), *Prehistoric Thrace. Proceedings of the international symposium in Stara Zagora* (pp. 161-171). Archaeological Institute with Museum and Museum of History, Stara Zagora. Sofia and Stara Zagora.
2004b The everyday life and the symbolism in the prehistoric Balkans. In *Criteria of symbolicity.* Retrieved from http://www.semioticon.com/virtuals/symbolicity/everyday [accessed 22.01.2017]

ROMANUCCI-ROSS, L.
1969 The Hierarchy of Resort in Curative Practice: The Admiralty Islands, Melanesia. *Journal of Health and Social Behvior* 10: 201-209.

SCARBOROUGH, J.
1969 *Roman medicine.* Ithaca, NY: Cornell University Press.

SCHIFFER, M. B.
1987 *Formation Processes of the Archaeological Record.* Albuquerque: University of New Mexico Press.

SEKIME, T. and KAMURO, H. (eds).
1998 *Djadovo excavation 1997. A preliminary report on the 11th excavation at Djadovo, Bulgaria.* Tokai University Thracian Expedition.
1999 *Djadovo excavation 1998. A preliminary report on the 12th excavation at Djadovo, Bulgaria.* Tokai University Thracian Expedition.
2000 *Djadovo excavations 1999. A preliminary report on the 13th excavation at Djadovo, Bulgaria.* Grant-in-Aid for Scientific Research (A), 1999 No. 09041032. Tokai University Thracian Expedition.

SMITH, A., BADALYAN, R. S., and AVETISYAN, P.
2009 *The foundations of research and regional survey in the Tsaghkahovit plain, Armenia.* Chicago: University of Chicago.

SOBO, E. J.
2004 Theoretical and applied issues in cross-cultural health research. In C. R. Ember and M. Ember (eds.), *Encyclopedia of medical anthropology: health and illness in the world's cultures* (pp. 3-11). Vol. New York: Springer.

WEISSHAAR, H.-J.
1996 *Die Deutschen Ausgrabungen Auf Der Pevkakia-Magula in Thessalien I: Das späte Neolithikum und das Chalkolithikum.* Mainz: R. Habelt.

Cyrillic

Катинчаров Р., Мерперт Н.Я., Титов В.С., Мацанова В.Х. & Авилова Л.И.
1995 *Селищна могила при село Юнаците (Пазарджишко).* Т. 1. София: Агато & Диос.

Recycling Power and Place:
The Many Lives of Traprain Law, SE Scotland

Ian Armit, Andrew Dunwell, Fraser Hunter

Abstract

Traprain Law, south-east Scotland, is best known for the spectacular hoard of Late Roman silverware recovered from the hillfort in 1919. Traditionally regarded as the capital of the philo-Roman Votadini, the hill has a much deeper history, revealed through several episodes of investigation from the early twentieth century onwards. Excavations from 1999–2006 were accompanied by a new programme of AMS dating which has greatly clarified the complex biography of Traprain Law, suggesting several major transformations from the Early Bronze Age to the Early Medieval period. During much of the Bronze Age and pre-Roman Iron Age, the hill seems to have been a place of special significance marked by the creation of extensive rock art panels, depositions of fine metalwork, and the construction of several enclosing 'ramparts'. During two key episodes, however, in the ninth century BC, and again in the Roman Iron Age, the hill seems to have become densely occupied, with signs of high status activity, before turning once again into a place of primarily religious significance in the Early Medieval period. This paper examines how far each of these episodes of human engagement with the hill drew upon earlier accretions of meaning.

The Archaeology of Traprain Law

Traprain Law is a large volcanic hill situated around 30 km east of Edinburgh in East Lothian, south-east Scotland (fig. 1). Dominating the coastal plain, it is one the largest hillforts in Scotland (fig. 2). Archaeologically, the site is probably best known for the spectacular hoard of Late Roman silverware, known as the 'Traprain Treasure', found there in 1919 (Curle 1923). The discovery came during the early excavations of the hill by Alexander Curle and James Cree (e.g. Curle 1920; Cree and Curle 1922). Since their work, which yielded an unparalleled amount of imported Roman material for an indigenous site in Scotland, Traprain Law has been regarded as the capital of the Votadini; an indigenous tribe named by Ptolemy, who were thought to have inhabited SE Scotland during the Roman period (e.g. Hogg 1951; Gillam 1961). The hill, however, has a much longer history, which has been progressively revealed through several episodes of excavation from the early 20th century onwards (Armit 2001).

The early excavations focused on an area of relatively level ground at the west end of the hill, just below the summit (fig. 3). This western shelf, or terrace, produced significant quantities of artefactual material, but very little information on the nature of the accompanying occupation deposits. Although numerous hearths were identified, Curle and Cree were unable to identify the remains of any buildings. This was not altogether surprising since walls of turf or timber would have been almost impossible to identify with the techniques current at the time.

Subsequent excavations have been carried out on the hill by a number of investigators including Gerhard Bersu (published posthumously in summary, Close-Brooks 1983), Stewart Cruden (1939), and Peter Strong (1984). Although Bersu carried out some small scale work on the summit, this later excavation has otherwise exclusively focused on aspects of the rampart system. In addition to this invasive work, an important survey of the defences by Richard Feachem (1955) still provides much of the framework for overall interpretation of the rampart system. A major summary of the site sequence was also published (Jobey 1976) and this, together with our own recent work, informs the interpretations given here.

Recent Work

Our work on the site began in 1999 with the Traprain Law Summit Project (TLSP) which sought to examine the poorly understood evidence for settlement and other activity on the extensive, but largely unexplored, summit area (Armit et al. 2002). This was designed to counteract the previous dominance of the western shelf in attempts to understand the history of activity on the hill. Then, in summer 2003, a devastating fire burnt out large areas of the hill, leading to a rescue excavation programme on the most badly damaged sectors. The Traprain Law Fire Damage excavations (TLFD) enabled us not only to elucidate further aspects of the summit area, but also to investigate some elements of the defences (Armit et al. 2006). Overall we have now excavated more than 20 trenches and numerous test-pits at different points across the hill.

A major aim of the recent work has been to obtain the first scientific dates from the site to overcome the previous reliance on artefact typology, and to explore some specific issues relating to the ebb and flow of human occupation of the hilltop (Armit, Dunwell and Hunter, in prep.). A programme of AMS radiocarbon dating has now enabled us to start unravelling the

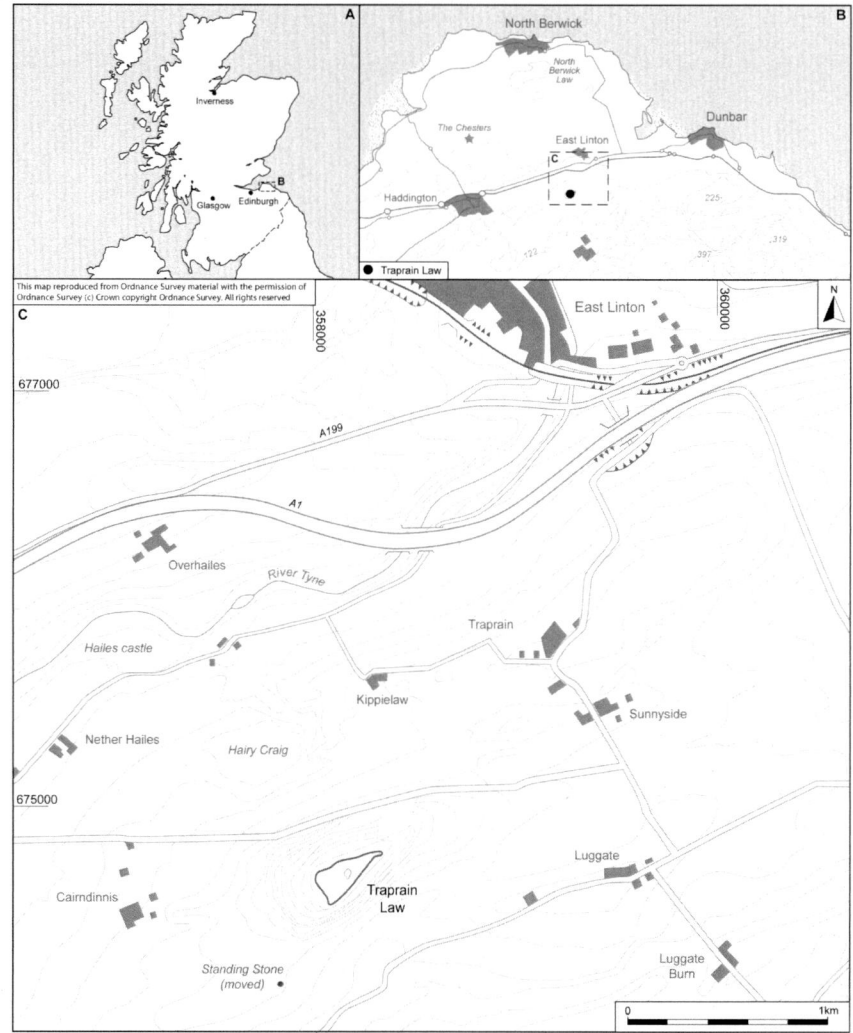

Fig. 1: *Location map (drawn by Rachael Kershaw)*

Fig. 2: *Traprain Law, East Lothian (photo: Ian Armit)*

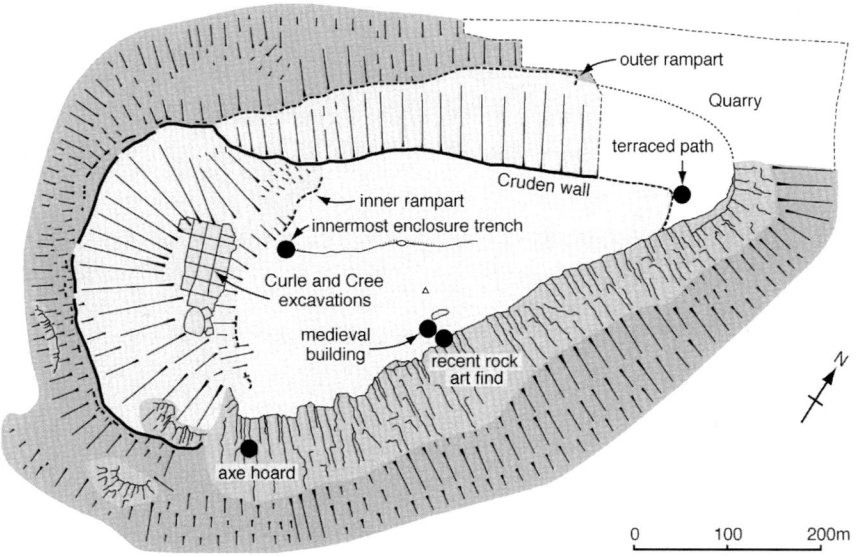

Fig. 3: *Simplified plan of Traprain Law showing the main focal areas of excavation (drawn by Libby Mulqueeny)*

complex chronology of the site. For the first time, we can begin to see the outlines of a long and apparently discontinuous history of occupation. The emerging pattern implies several major transformations in the ways in which the hill was perceived and inhabited from the Early Bronze Age to the Early Medieval period. Crucially, each of these transformations involved appropriating past perceptions of the hill – recycling its past associations to build new meanings and new relationships.

The remainder of this paper has two main parts: firstly, we will outline the conventional narrative based primarily on the early excavations by Curle and Cree and others. We will then examine the new sequence emerging from the recent excavations and, especially, from the AMS dating programme, in relation to the concept of recycling and reuse.

The Conventional Narrative

The conventional narrative for Traprain Law sets out a broadly unilinear trajectory of growth leading up the establishment of Votadinian power in the Roman Iron Age and the subsequent transmutation of the Votadini into the kingdom of Gododdin in the early centuries AD. Feachem, for example, saw a progressive expansion from an original 10 acre summit enclosure through to a 40 acre 'oppidum' in the late pre-Roman Iron Age, with some final retrenchment to a 30 acre version in the late Roman period (Feachem 1955, see especially figure 4).

In broad terms, the conventional sequence can be divided into 4 phases:

1. *Early activity*: the early excavations produced a scatter of Mesolithic and Neolithic flints and polished stone axes at various locations on the hill. They also recovered a small number of Middle Bronze Age urned cremation burials, suggesting that the Law may have been used as a place of burial at this time. Quarrying operations in the 1930s uncovered some remarkable rock art (see below) but this has figured little in previous interpretations. Overall, this early activity was seen as essentially a series of disparate and unconnected visitations largely irrelevant to the later use of the hill.

2. *The Late Bronze Age*: it was clear from Curle and Cree's excavations that there was significant Late Bronze Age activity on the western shelf of Traprain Law. Bronze objects, along with moulds for axe, sword and spear-head production, suggested occupation during the Ewart Park phase, originally placed around the 7th – 6th centuries BC (e.g. Coles 1960) but which would now be dated to around 950-700 BC (Needham *et al* 1997: 93). But there was always a question as to whether these remains represented, on the one hand, a restricted area of activity on the western shelf, or on the other, a Bronze Age origin for the hillfort as a whole. Given the conservatism of hillfort dating in the UK, archaeologists have been generally reluctant to accept a Bronze Age date for the defences. An isolated Bronze Age settlement on the western shelf was therefore postulated (e.g. Jobey 1976: 193). Occasional objects, such as an early iron axehead suggested that this settlement overlapped with the earliest adoption of an iron technology in the region (Jobey 1976: 195).

3. *The pre-Roman Iron Age*: Although the early excavations found very little artefactual material that could be definitively related to

pre-Roman Iron Age activity, successive writers have been at pains to fill this chronological gap (e.g. Jobey 1976: 195, supported by Hill 1987: 86). The conventional narrative thus proposed that the various lines of rampart that enclose the summit should belong to the Iron Age. There are perhaps three main reasons for this:

   i. Firstly, British hillforts are seen traditionally as an Iron Age phenomenon, and there has been resistance to suggestions that hillfort ramparts were constructed during the Bronze Age.
   ii. Secondly, the idea of Traprain Law as the pre–Roman capital of the Votadini (one of Feachem's Scottish 'oppida' (1955: 288-9)) required a pre-Roman origin for the site.
   iii. And thirdly, the pre-Roman Iron Age in Scotland is very finds-poor (virtually aceramic in many areas, and lacking in diagnostic metalwork); most hillforts, therefore, produce few artefacts that could have offer definitive dating evidence. In this context, a materially invisible Iron Age for Traprain Law, becomes more or less acceptable.

4. *The Votadinian capital:* Finally of course, the substantial quantities of Roman pottery and metalwork, and the deep deposits on the western shelf of the hill, combined to suggest a 'boom period' for Traprain in Roman period. The lack of Roman military installations in East Lothian (the nearest fort just over 20 km to the west at Inveresk on the eastern outskirts of Edinburgh), and the apparent survival of the ramparts at Traprain Law into the Roman period, also led to suggestions that the Votadini were allied to Rome; perhaps even a client kingdom. Despite some possible fluctuations in the supply of Roman material (not unexpected given the periodic movements of the northern frontier which saw Traprain Law as variously inside and outside of the imperial boundaries), this apparent prosperity lasted throughout the Roman period until the beginning of the 5th century AD. The subsequent destabilization of the Roman province and the withdrawal of Roman military support, in the traditional model, undermined Votadinian power causing the site to be abandoned. The latest fortification (known as the 'Cruden Wall') and the burial of the 'Traprain treasure' itself marked the last phase of occupation.

This traditional model, rendered in a rather simplified form here, did not go completely unchallenged. There was always some unease, for example, around the paucity of pre-Roman Iron Age evidence, and there were also suggestions that the Roman activity may actually have been ritual in character rather than signalling a secular capital (Hill 1987, though this view was heavily criticized at the time, Close Brooks 1987). But the basic picture was well established.

New Interpretations

The conventional narrative for Traprain assumes, after episodic visits in early periods, a continuous process of growth and gradual intensification from the Late Bronze Age onwards, until the site finally emerges, many centuries later, as the capital of the dominant tribe in southern Scotland (fig. 3). As such, it reflects the broader interpretive sequence for southern Scottish hillforts first established by C. M. Piggott in her excavations at Hownam Rings, Roxburghshire, in the Scottish Borders (Piggott 1948). The 'Hownam sequence' envisaged a progression from palisaded enclosures to univallate hillforts, thence to multivallate defences; finally, defences were abandoned altogether in the Roman Iron Age. The sequence was identified first at Hownam Rings and other sites in the Borders, and subsequently generalized as a model for southern Scotland. This powerful model only began to be dispelled during the late 1970s, after the application of open area excavation at sites like Broxmouth hillfort (Hill 1982), and Dryburn Bridge (Dunwell 2007), both in East Lothian. The influence of Piggott's model hangs heavily, however, over previous interpretations of Traprain Law (cf. Armit 1999).

Our recent work casts doubt on the established idea of continuity and progress at Traprain Law. Instead, recycling and reinvention seem to have been key factors for the various generations who inhabited the hill, building on the symbolic and material dimensions of the past in the creation of social and political identities. We will focus here on three key parts of the sequence where the recent work has had most impact: the rock art; the Bronze Age occupation; and the emergence of the Iron Age capital.

Rock Art

The first sign of Traprain's regional importance is marked by the creation of a series of rock carvings, probably begun during the Late Neolithic period and carrying on into the Early Bronze Age. The largest panels were found on the north-east side of the hill during quarrying in the 1930s (Edwards 1935). Despite some basic recording, however, these carvings were destroyed by blasting shortly after their discovery in order to allow the quarry to progress. As they were lost from the landscape at a time when archaeologists generally took very little interest in rock art, these carvings have seldom been considered in relation to the wider development of human activity on the Law. Examination of casts and surviving fragments shows that the first carvings were dominated by cup-and-ring marks (McCartney 2003). Subsequently, however, these fairly conventional pecked motifs were overlain by a mass of incised, linear carvings forming complex designs. The resultant panels were by far the largest and most elaborate in SE Scotland, and they have no obvious parallels in the Atlantic European rock art tradition. When combined with the physical prominence of Traprain Law as a location, the creation of these carvings was clearly highly significant. Traprain had always been a dominant presence in the landscape and now it became inscribed with elaborate and extensive carvings which, although we cannot now decipher them, were full of meaning nonetheless. As we shall see below, this delineation of a 'meaningful' Traprain was to be referenced by future generations.

The Late Bronze Age

Perhaps the main impact of the recent work at Traprain Law has been in relation to the Bronze Age occupation. During excavation, it was already clear that prehistoric occupation deposits were present in almost every trench. Since the ceramics of the southern Scottish Bronze and Iron Ages are at least superficially rather homogenous, however, it was difficult to be more precise in assigning a date to this material in the field. An integral part of our research, however, was a programme of AMS (Accelerator Mass Spectrometry) radiocarbon dating. The process of sample selection was extremely rigorous in only using contexts where there was no evidence for later soil movement (the deposition of colluvial deposits is a major problem in interpreting the archaeology of the hill). Preferred samples were individual cereal grains from well-sealed contexts. Wherever possible three AMS dates were obtained from each context to identify any possible mixing of deposits and to add security of interpretation. Inevitably, this approach limited the number of potential samples, but the benefit is that the resultant dates provide a robust picture of human activity on the hill. Our main aim was to identify the date range for the major periods of settlement activity on the hillfort – partly to clarify the extent of Bronze Age activity, and partly to find the 'missing Iron Age' (should it exist). For this reason, we avoided any contexts containing Roman material, on the basis that AMS dating would be unlikely to provide more precise dates than the typological dating of the artefactual material.

The AMS results were very striking (fig. 4). Of the 36 dates received so far, from locations all across the hilltop, all but one date to the Late Bronze Age: the great majority focus on the 10th – 9th centuries BC, while others reflect activity in the later 2nd millennium BC. There appears to be a cessation of this intense human activity some time before 800 BC (the iron axehead from the earlier excavations notwithstanding).

The nature of the bronzework made and used on the site, together with the appropriation of such a dominant natural location and the extent of the excavated deposits, suggests that Traprain Law was an elite centre during the Late Bronze Age. This impression is supported by the discovery in 2004 of a votive axe hoard (fig. 5) which contained non-local axes, including examples probably from as far afield as Ireland and Yorkshire (Cowie pers. comm.), and by the rare occurrence of gold on the summit.

As part of this Bronze Age appropriation of the hilltop, specific attempts5seem to have been made to recycle elements of the Law's earlier religious significance. In the excavations of 2004, a previously unknown rock art panel was uncovered on an outcrop above the cliffs on the south edge of the hill (fig. 6). This had been incorporated into the floor of a Late Bronze Age building where the motifs would have been visible next to a substantially built hearth. What had originally been an open-air, 'public' monument had thus been appropriated within a closed and roofed structure, limiting and controlling access; part of a wider assertion of control by the elite based on Traprain. It is unfortunate that the circumstances in which the previously known rock carvings were discovered in the 1930s did not allow any archaeological exploration of their context – were they too enclosed during the late Bronze Age occupation?

A further piece of rock art, this time a fragment of the linear designs, was also found, redeposited in the fabric of a medieval building (Rees and Hunter 2000) close to the summit, and close to where the new panel was discovered in 2004. This piece appears seemed to have been deliberately squared-off and turned into a small plaque (fig. 7); it is impossible to date the period of its manufacture, but it is not impossible that it occurred during the Late Bronze Age when there was intense activity on that part of the site.

In two quite separate ways we see the recycling of the Traprain rock art within wholly new contexts where they would have taken on new meanings; the first transformed from open, landscape feature to closed private possession; the second recycled from landscape feature to artefact. It is tempting to see these transformations in relation to the enhanced status of the Late Bronze Age settlement, with spiritual power being recycled to legitimate a secular authority; the power of place recycled as the power of the social elite.

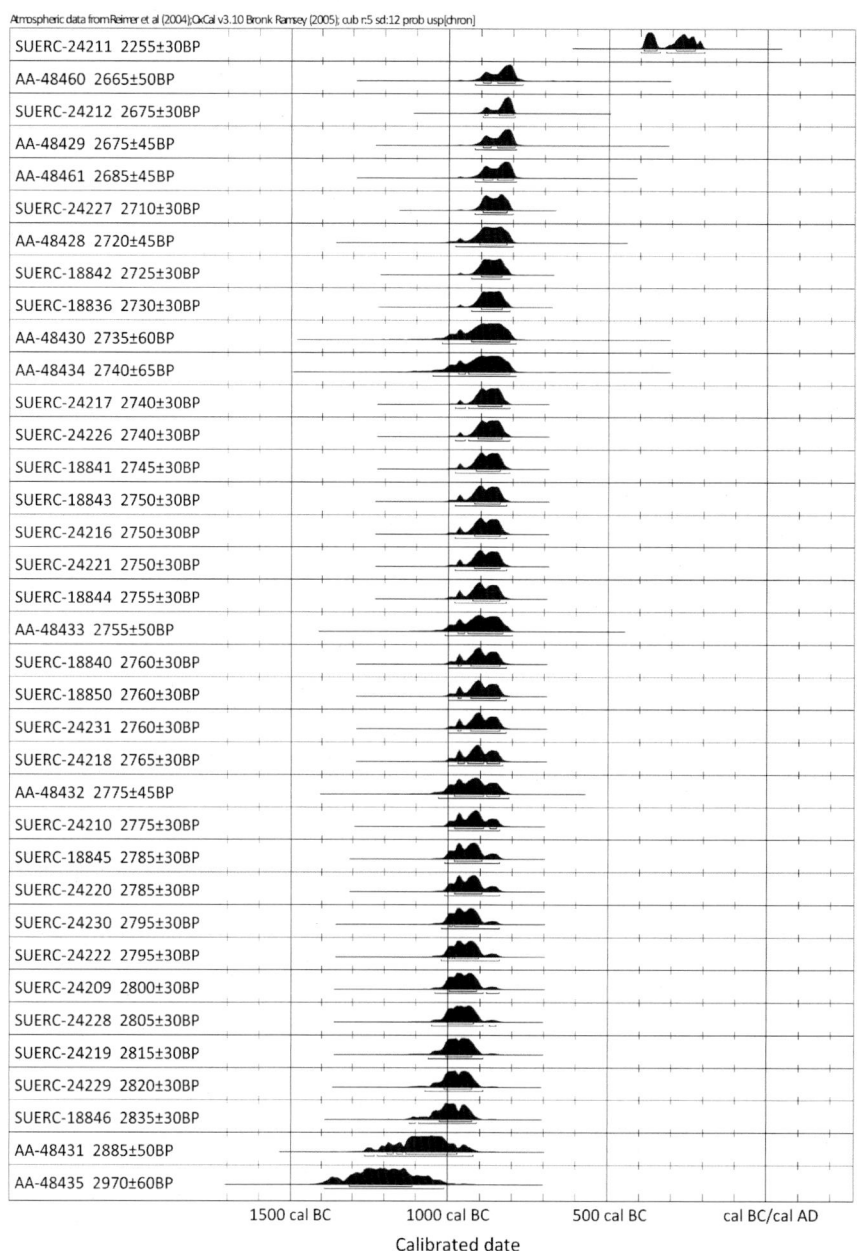

Fig. 4: *AMS dates from Traprain Law, funded by Historic Scotland, calibrated using Oxcal 4 (Bronk Ramsey 2009, Reimer et al. 2004)*

Recycling of the rock art was not the only way in which the Late Bronze Age elite attempted to legitimate their power at Traprain. The wider landscape was also drawn into the layout and construction of the settlement. An enclosure bank which defines part of the summit area and which is apparently dated to the Late Bronze Age (based on AMS dating) has a single out-turned entrance which frames with nearby hill of North Berwick Law (fig 8); another volcanic plug which, like Traprain Law, bears surface traces of prehistoric settlement and enclosure. Recent excavations in the vicinity of Traprain have demonstrated that Late Bronze Age enclosure is not necessarily unique or precocious on a regional scale (Haselgrove 2009).

Other entrances at Traprain are similarly directed towards other prominent natural features like the Bass Rock and Arthur's Seat; all volcanic hills that structure the regional landscape and would undoubtedly have accumulated mythological or cosmological associations by the later prehistoric period, as they had in later times. The name of Arthur's Seat in Edinburgh, for example, references Arthurian mythology, while Traprain Law itself acquired associations with important saints and mythological figures such as the eponymous King Loth from whom Lothian drew its name (referenced for example in Jocelyn's *Life of St Kentigern*). It is unlikely that any clear distinctions would have been drawn between such natural features and the man-

Fig. 5: *The axe hoard found in the burnt out area of Traprain Law in 2004 (National Museums of Scotland)*

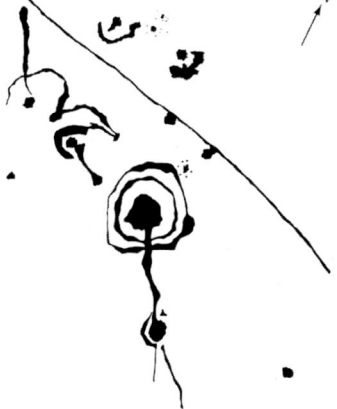

Fig. 6: *The axe hoard found in the burnt out area of Traprain Law in 2004 (National Museums of Scotland)*

made creations such as the rock art which adorned Traprain Law. All were cultural resources from a remote past which could be drawn together at Traprain Law in the creation of social power in the present.

Iron Age

It is clear from the spread of AMS dates that there was no widespread settlement across the summit of Traprain Law in the pre-Roman Iron Age. It may yet be the case that some deposits which contained Roman pottery (and thus were deliberately not selected for the initial dating programme) may have originated in the pre-Roman Iron Age; there are some objects from the early excavations, such stone balls and certain pottery forms, that may suggest some Iron Age presence. Yet if there was any activity on the hill between around 800 BC and the 1st century AD, it was seemingly on a far smaller scale than anything seen in the Late Bronze Age. It is possible that the multiple lines of rampart which enclose the hill were constructed in these centuries, though this remains to be demonstrated by AMS dating; in any case, this would point to activity of a rather different character to that seen in the Late

Fig. 7: *Stone 'plaque' made from fragment of the linear rock art (National Museums of Scotland)*

Fig. 8: *North Berwick Law framed in the out-turned entrance-way to the innermost enclosure at Traprain Law (photo: Ian Armit)*

Bronze Age, with the periodic enclosure of an essentially empty hilltop; a pattern seen elsewhere in the Scottish Iron Age, for example at the Brown Caterthun in Angus (Dunwell and Strachan 2007). Whatever the date of their construction, it appears that the centre of population which appears on the hill in the Roman Iron Age did not emerge from a long-occupied native hillfort. Indeed, where the relationship has been examined, it appears that the ramparts had been long abandoned before the Roman Iron Age occupation commenced (Armit *et al.* 2002).

Recycling and Renewal

Far from representing the expansion of an existing hillfort, the establishment of the Roman Iron Age settlement on Traprain Law (the Votadinian capital, if such it was), was effectively a new foundation on a long abandoned but well-remembered hill. If the Votadini were **not** a long-established authority in the region, but essentially a creation of the Roman period, then the choice of Traprain Law for their capital once again shows the importance of the symbolic power of the past as a strategy to underpin power in the present. Votadinian power, especially if supported by Rome, would have required legitimation. That appears to have been attempted through the reclamation of a hilltop settlement that would have had strong associations with power and authority linked to a past now situated in mythological time (Gosden and Lock 1998). This past was still visible in the lines of collapsed rampart, the stances of old house foundations, and the panels of ancient rock art, but more importantly perhaps it was a past embedded in memories, tales and traditions of a (by now mythic) Bronze Age elite.

Throughout its history, each new appropriation of Traprain Law drew on the affordances of its deep and distant past, re-working the physical remnants of past occupations, and draping the hill with fresh layers of meaning. In its many lives, each separated by several centuries, the recycling of the past remained a constant and unifying theme.

Acknowledgements

The recent excavations at Traprain Law were funded by Historic Scotland, the National Museums of Scotland, the Society of Antiquaries of Scotland, the Russell Trust and the Munro Lectureship Trust.

Bibliography

ARMIT, I.
  1999 Life after Hownam: the Iron Age in South-East Scotland, pp. 65-79, in Bevan, B. (ed.) *Northern Exposure: Interpretative Devolution and the Iron Ages in Britain.* Leicester: Leicester Archaeology Monographs, No. 4.
  2001 'Traprain Law', *British Archaeology* 57: 8-11.
ARMIT, I., DUNWELL, A. J. and HUNTER, F.
  2002 The hill at the Empire's edge: recent work on Traprain Law. *Transactions of the East Lothian Antiquarian and Field Naturalists' Society*, XXV: 1-11.
  in prep. *The Hill at the Empire's Edge: Excavations on Traprain Law 1999-2006.* Edinburgh: Society of Antiquaries of Scotland, Monograph Series.
ARMIT, I., DUNWELL, A. J., HUNTER, F. J., McCARTNEY, M. and NELIS, E.
  2006 Traprain Law. *Current Archaeology* 203: 602-7.
BRONK RAMSEY, C.
  2009 Bayesian analysis of radiocarbon dates, *Radiocarbon*, 51.1: 337-60.
CLOSE-BROOKS, J.
  1983 'Dr Bersu's excavations at Traprain Law, 1947', pp. 206-223. In O'Connor, A. and Clarke, D. V. (eds.) *From the Stone Age to the 'Forty-Five.* Edinburgh: John Donald.
  1987 'Comment on Traprain Law', *Scott Archaeol Rev* 4: 92-4.
COLES, J.
  1960 Scottish Late Bronze Age metalwork: typology, distributions and chronology. *Proceedings of the Society of Antiquaries of Scotland* 93: 16-134.

CREE, J. E. and CURLE, A. O.
  1922 'Account of the excavations of Traprain Law during the summer of 1921', *Proceedings of the Society of Antiquaries of Scotland* 56: 189-259.

CURLE, A. O.
  1920 'Report of the excavation on Traprain Law in the summer of 1919', *Proceedings of the Society of Antiquaries of Scotland* 54: 54-124.
  1923 *The Treasure of Traprain*. Glasgow.

DUNWELL, A.
  2007 *Cist burials and an Iron Age settlement at Dryburn Bridge, Innerwick, East Lothian.* www.sair.org.uk. Scottish Archaeological Internet Report 24.

DUNWELL, A. and STRACHAN, R.
  2007 *Excavations at Brown Caterthun and White Caterthun Hillforts, Angus, 1995-1997.* Tayside and Fife Archaeological Committee Monograph 5, Perth.

EDWARDS, A. J. H.
  1935 'Rock sculpturings on Traprain Law, East Lothian', *Proceedings of the Society of Antiquaries of Scotland* 69: 122-37.

FEACHEM, R. W.
  1955 'The fortifications on Traprain Law', *Proceedings of the Society of Antiquaries of Scotland* 89: 284-9.

GILLAM, J. P.
  1961 'Roman and native, AD 122-197', pp. 60-90, in Richmond, I. A. (ed.), *Roman and Native in North Britain*. London: Thomas Nelson and Son..

GOSDEN, C. and LOCK, G.
  1998 Prehistoric Histories. *World Archaeology* 30.1: 2-12.

HASELGROVE, C. C.
  2009 *The Traprain Law Environs Project: fieldwork and excavations 2000-2004.* Society of Antiquaries of Scotland. Edinburgh.

HILL, P. H.
  1982 Broxmouth Hillfort Excavations 1977-1978, an interim report. In Harding, D. W. (ed.) *Later Prehistoric Settlement in South-East Scotland.* University of Edinburgh Department of Archaeology Occasional Paper No. 8, Edinburgh, pp. 141-188.

HILL, P.
  1987 'Traprain Law: the Votadini and the Romans', *Scott Archaeol Rev* 4, 85-91.

HOGG, A. H. A.
  1951 'The Votadini', pp. 200-21, in Grimes, W. F. (ed.), *Aspects of Archaeology in Britain and Beyond: Essays Presented to OGS Crawford.* London: H. W. Edwards.

JOBEY, G.
  1976 'Traprain Law: a summary', in Harding, D. W. (ed.) *Hillforts: later prehistoric earthworks in Britain and Ireland*, 191-204. London: Academic Press.

McCARTNEY, M.
  2003 *Rock Carvings from Traprain Law.* Unpublished undergraduate thesis. Queen's University Belfast.

NEEDHAM, S., BRONK RAMSEY, C., COOMBS, D., CARTWRIGHT, C. and PETTITT, P. B.
  1997 An independent chronology for British Bronze Age metalwork: the results of the Oxford Radiocarbon Accelerator Programme. *Archaeological Journal* 154: 55-107

PIGGOTT, C. M.
  1948 'Excavations at Hownam Rings, Roxburghshire, 1948'. *Proceedings of the Society of Antiquaries of Scotland* 82: 193-225.

REES, T. and HUNTER, F. J.
  2000 'Archaeological excavations of a medieval structure and an assemblage of prehistoric artefacts from the summit of Traprain Law, East Lothian, 1996-7', *Proceedings of the Society of Antiquaries of Scotland* 130: 413-40.

REIMER, P. J., BAILLIE, G. L., BARD, E., BAYLISS, A., WARREN BECK, J., BERTRAND, C. J. H., BLACKWELL, P. G., BUCK, C. E., BURR, G. S., CUTLER, K. B., DAMON, P. E., EDWARDS, R. L., FAIRBANKS, R. G., FRIEDRICH, M., GUILDERSON, T. P., HOGG, A. G., HUGEN, K. A., KROMER, B., Mc.CORMAC, G., MANNING, S., BRONK RAMSEY, C., REIMER, R. W., REMMELE, S., SOUTHON, J. R., STUIVER, M., TALAMO, S., TAYLOR, F. W., van der PLICHT, J., WEIHENMEYER, C. E.
  2004 IntCal04 Terrestrial Radiocarbon Age Calibration from 0–26 cal kyr BP. Radiocarbon 46.3: 1029-1059.

STRONG, P.
  1984 *Excavation of a possible rampart at Traprain Law, East Lothian*, Unpublished interim report, Scottish Development Department (Ancient Monuments).

Tells as Recycled Places.
Experimenting the Chalcolithic Ritual Technologies of Construction and Deconstruction

Dragoş Gheorghiu

Abstract

The Chalcolithic *tell*-settlements from Eastern Europe were the result of a series of standardised and repetitive actions of construction and deconstruction, the old materials incorporated into the new dwellings creating a syncretism between technology, symbolism, and ritual.

Motto:

>Didst thou hap to see
>Somewhere down the lea
>An old wall all rotten,
>Unfinished, forgotten? [...]
>As the prince did hear
>Greatly did he cheer,
>And walked to that wall,
>"Here's my wall!" quoth he.
>"Here I choose that ye
>Build for me a shrine,
>A cloister divine.
>
>(Mesterul Manole /The Ballad of Master Manole)

Introduction

The motto of the present paper belongs to a ballad, which circulated in the Middle Ages and Post-Medieval period in the Balkans, describing the epopee of a group of master masons employed by a certain Black King to search for a ruin, to build a monastery on top of it. Although being a historical folk production, the ballad about the recycling of a place depicted an ancient Balkan custom, dating back to the emergence of prehistoric *tell* settlements. The present paper will approach the subject of recycling places in the prehistoric north Balkans – Lower Danube area, using experimental archaeology as an instrument to approach ancient ritual experience.

Tell settlements emerged in the Near East as a strategy of long term settling on the very surface of the land, to cite only the occupation at Çayönü (Özdoğan and Başgelen 1999), or at Yümük Tepe (Caneva 1999). In the Balkans and the Lower Danube area the tell settlements, which became frequent during the Chalcolithic period, did not last so long and had smaller dimensions (see Todorova 1982). Scholars underlined the trait of the compactness (Sherratt 1983; Chapman 1990) of this type of settlement, and the large quantities of clay employed to raise the buildings (Sherratt 1983; see also Kalicz and Raczky 1990). Their definitions of *tells* therefore placed stress on the cumulative character of the cultural and natural processes of site formation, perceiving the settlement as being an artificial clay geomorph, or artefact (Gheorghiu 2013). Clay, as a recyclable (Hurcombe 2007: 184), eco-friendly, and economical material (Minke 2006; Reddy 2007), was used intensively for building in the loess or alluvial areas, because of its easy extraction, preparation and possibility of creating composite and reinforced materials, its biodegradability, its high thermal comfort and its potential for developing repair or expansion operations of certain existing constructive elements (Minke 2006; Creangă et al. 2010).

Tell-settlements

If in Southern Europe the emergence of *tells* dates back to the 6th millennium BC (Whittle 1996; Perlès 2001), in the Lower Danube area they become manifest centuries later (Comsa 1997; Neagu 1997; Bailey 1997). Although *tells* were not the only form of settlement in this region, a growing number of data show that at the beginning of Gumelniţa –Karanovo –Kodzadermen (see Comşa 1990; Bailey et al. 2000; Ştefan 2010), they were *part* of a complex system of habitation, based on a mixed economy and an intermittent occupation (see Dumitrescu 1986; Bailey 1997: 52; Erdoğu 2005: 31) of some "strategic" (see Morintz 1962; Morintz 2007: 36) places. These examples tend to present the *tell* settlement as a sort of seasonal place of refuge, for protecting people and animals, and a storage space during the cold season.

A frequent pattern of spatial organisation in the Lower Danube Chalcolithic was as follows: the settlement was positioned near water, on a geomorph situated above the flood level (Gheorghiu 2006a), with the entrance facing south and with good visibility towards the cultivated land (Gheorghiu 2003). A separation from the mainland was achieved with ditches and palisades (Comşa 1990), as suggested by the reconstruction in Fig. 1.

To live in a *tell* settlement was a spatial experience fundamentally different from that offered by the previous Neolithic settlements, because it compelled to new sets of behavioural rules. The compact shape of the inhabited space was the result of the building of a large number of wattle-and-daub houses on a minimal soil surface, due to geographic restrictions and safety reasons (for a discussion of the relationship between built and un-built spaces in a *tell* settlement see Chapman 1990; 1991a; 1991b). In the present paper I would try to approach only one specific aspect of *tell* settlements, that of the recycling of place and materials, using experimental archaeology as the main instrument of investigation.

Tell-settlements as Places of Recycling. Experiments of Construction and Deconstruction

The reconstruction of the wattle-and-daub prehistoric structures initially involved the study of the ethnographic processes of building and recycling in traditional dwellings. For example, I examined how a traditional household in the village Uzunu (Giurgiu County) recycled the material excavated from a nearby Chalcolithic tell, which had functioned as clay source for the village for a century (Fig. 2), or how the fishermen from Hârşova (Constanţa County) used the material from the local Chalcolithic tell (Fig. 3).

Another source of information was the experiments with house building I carried the last decade which offered data about the behaviour of building materials and about the operations of design necessary to start the building process. In this perspective, one can infer that the first technological stage of the building of a new settlement was the procurement of the ligneous material for palisade and house building, which needed to be cut, transported and then kept to dry, a long time before the beginning of the first constructions. A second stage was the tracing of the plan of all the houses on the soil's surface, followed by the perimeter of the ditch and palisades (Gheorghiu 2009a). The dimensions of the perimeter ditches created debates about their efficiency for protection (Marinescu-Bîlcu *et al.* 1997), experiments inferring for them a role of clay source and a place for preparing the wet clay mixed with straws for the palisade and the houses.

Fig. 1: *The experimental reconstruction by the author of the first level of dwelling in a Chalcolithic settlement (Vădastra 2003).*

The emergence of *tell*-settlements in the Near East and South Europe was possible due to the extensive use of clay as building material. When mixed with water and straws, clay transformed itself into a plastic composite material with insulation properties (see Minke 2006) (Fig.4). To increase the elasticity of the material resulted, and to confer a higher resistance to water and a higher stability, diverse recycled organic substances like dung, blood, urine, casein, or whey were [probably] added (see Minke 2006). From the list of organic materials I identified in some cases the use of dung in the clay used for interior architecture

Fig. 2: *Uzunu tell used as a clay quarry by villagers (2007).*

Fig. 3: *Fishermen houses at the base of Hârşova tell (2005).*

Fig. 4: *The building of a wattle and daub house (2003).*

Fig. 5: *Fragment of a burned wall with visible vegetal straws. (Sultana tell).*

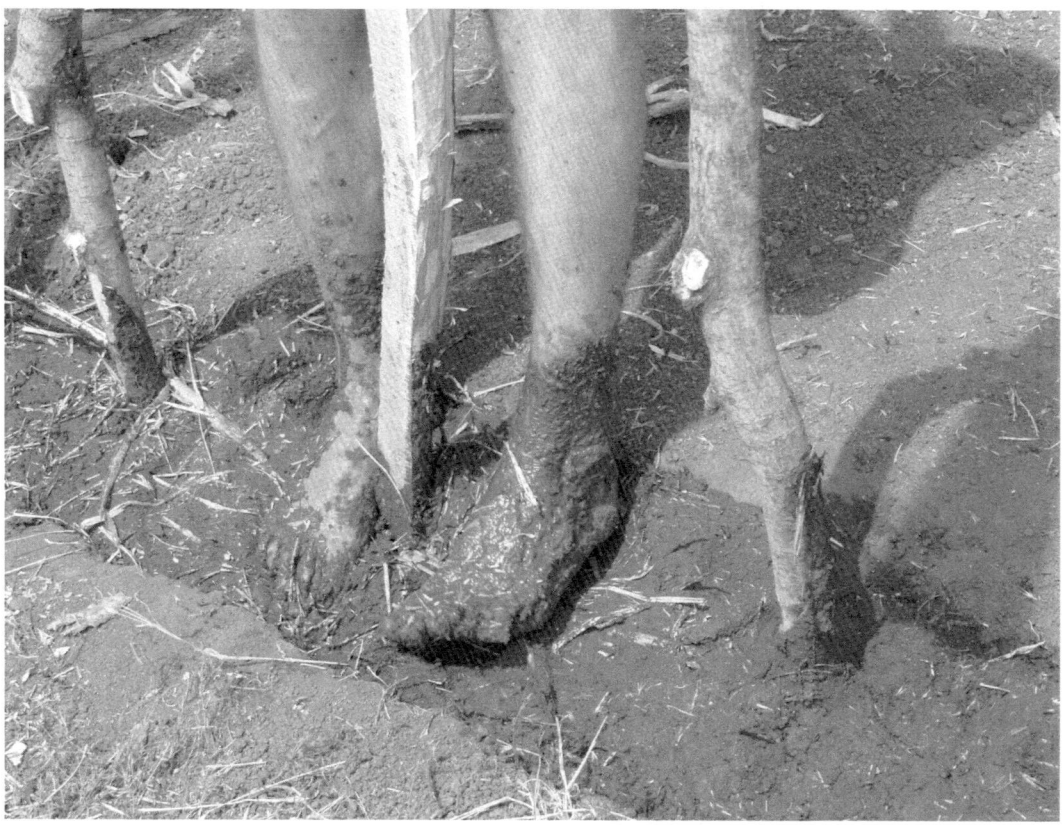

Fig. 6: *A foundation trench (Vădastra experiments 2003).*

features. An indirect evidence of the presence of this organic material is the standardized dimension of the vegetal fragments, which can be seen in the fragments of the burned walls. Their length is the same with the vegetal fragments to be found in cattle dung. (Fig. 5)

For a good protection of the surface of the daub walls or floors, a cyclical process of plastering (Minoiu *et al.* 2005) with clay mixed with organic substances was necessary (as demonstrated by contemporary ethnographic models, Ghinoiu 2003; Hadjiri *et al.* 2007; Creanga *et al.* 2010), this operation permitting archaeologists to approximate the duration of life of a building. Both straws and animal dejections which were added to clay were recycled sub-products of the Chalcolithic agro-pastoral economy, and the operations of building and plastering could be seen as a first stage of the process of recycling in a *tell* settlement.

Since clay buildings without an internal armature have a lower resistance to earthquakes (Vargas *et al.* 2005; Minke 2006), the Chalcolithic builders employed an efficient method to fix in the soil the vertical wooden structure and the wattle (Gheorghiu 2007a; 2009a), using foundation trenches (Gheorghiu 2005; Gheorghiu 2007a; 2006b; 2016). A foundation trench with a "U" or "V" shaped profile and a depth of 80-90 cm (see http://www.culture.gouv.fr/fr/arcnat/harsova/ro/f-plan.htm; accessed 22.01.2017) permitted the thrust of the ligneous material on its bottom and an additional fixing of it by filling the void with pressed soil. (Fig. 6) The relevance of the foundation trenches for the understanding of the Chalcolithic process of recycling is crucial, since their making recycled the material of the lower level of dwelling in a *tell* settlement.

After a phase of intense dwelling the settlement was deserted for a period of time (see Haita 1997: 88; Haita 2000: 53). In many cases, when the built features of the settlement were not intentionally demolished (Marinescu-Bilcu *et al.* 1997: 66), the deserted houses were intentionally burned down (see Haita 1997: 88), a cultural trait frequent in Chalcolithic South Eastern Europe (Tringham 1992; Tringham and Kristič 1990; Stevanovič 1997; 2002; Chapman 1999), which I approached from the perspective of experimental archaeology (Gheorghiu 2005; 2007a; 2007b; 2014; 2016).

The experiments of intentional combustion of a wattle and daub house (8 x 3.5 x 2.3 m, an average dimension of a house in a Gumelnița tradition *tell*), filled with textiles, wooden furniture, cereals, and a wooden pile for the oven, produced a material record similar with the one identified in the archaeological excavation.

After several hours of intense combustion, with a strong air-draught produced by the channels created after the total consumption of the ligneous material of the posts deeply included in the walls' clay (Gheorghiu 2005; Gheorghiu 2016), one can observe that combustion did not affect completely the built forms: alongside the parts transformed completely into ceramic material (Fig. 7), there were also places where the effect of fire was minimal, to cite only the unburned wood posts preserved on the surface of the soil (Fig. 8) made visible after the collapse of the walls inside the built perimeter. Due to the high effort needed to produce the wood pieces for building, one can infer that a large part of this unburned ligneous material was possible to be recovered post burning. Because of the internal structure of the house (like separating walls, posts and clay pyrostructures) the collapse of the walls could not create an even, homogenous surface, even after years of weathering. (Fig. 9)

Fig. 7: *The ceramic wall of a burned house (Vădastra 2006).*

Fig. 8: *Unburned part of a structural post (Vădastra 2010).*

Fig. 9: *The burned down house three years after the collapse (Vădastra 2009).*

As experiments showed, a post-abandon landscape five years after the burning down of a house was a "wild" one: the borders of the perimeter ditch collapsed, reducing their depth to a half, the volume of the debris also diminished, and was totally covered by vegetation.

In this instance the settlers returning to start to live on the tell were compelled to squash the debris of the previous dwelling phase (Marinescu-Bilcu *et al.* 1997: 66; Popovici *et al.* 2000: 17).

The following operation was the excavation of the foundation trenches, positioned in such a way not to overlap completely the previous houses' perimeter, because the loose soil could not offer good stability for new buildings (Gheorghiu 2002: 99ff; for Cucuteni tradition see Ursulescu and Tencariu 2002). This first stage of reoccupation of a surface already dwelt determines the new spatial organization of the second level of dwelling in a tell settlement, therefore several repositions due to several overlapped levels of

Fig. 10: *An unburned wattle and daub house left eight years to weathering with a recyclable wooden structure (Vădastra 2011).*

Fig. 11: *A mass of large burned architectural fragments (Uzunu tell, 2007)*

occupation on the same surface of land will produce a change in the orientation of numerous buildings (Gheorghiu 2002: 99, fig. 6.10). Traces of recycling are visible everywhere on Level 2: the space between the palisades was filled with soil and rubbish from the destroyed houses (Todorova 1978: 49), the old ditches were filled with the same material (Comşa 1986: 61), and the floors were prepared from crushed walls (Marinescu-Bilcu 1997: 69).

An immediate source for ligneous material could have been also the unburned deserted houses whose clay fabric was washed up by weathering. (Fig. 10) If a *tell* settlement was not completely burned down at the moment of its abandonment, the people returning to resettle the place could find there dry wood ready to be reused.

Since the processes of construction, deconstruction and recycling took place on a specific perimeter, restricted by ditches and palisades, the result was a growth of the volume of the *tell* settlement, and a recycling of some of the old built features (Fig. 11); for example parts of the palisade were transformed into walls for the peripheral houses (see Todorova 1982: 212, fig. 165).

After many episodes of dwelling, a tell settlement will look as a multi-layered cake with layers rolled up, every new layer containing recycled parts from the old ones (see Comşa 1986: 66).

Chaîne-opératoire, Ritual and the Rituality of Technology

As experimentalist I was fascinated by the technological process of building and recycling which functioned during the (discontinuous) existence of a *tell* settlement, and tried to identify and experience the ritual stages of this mode of dwelling, since rituality was a main characteristic of prehistoric societies (see Bradley 2005). After more than a century of Modernity, scholars are no more accustomed to perceive the social dimension of ancient rituality and the study of tells is one example, since it involves a technological ritual

The complex technological process of construction, deconstruction and recycling had a tripartite ritual structure, specific for the *rites of passage* (van Gennepp 1960 (1908); Turner 1995), and functioned as follows: the utilization of an object, a material or a place was followed by a *liminal* stage of discard of materials and abandon of the place, which, at its turn was followed by a stage of *incorporation*. "Incorporation" could mean the reutilization of old objects, materials or places, or their physical inclusion into new materials, instances of both being found in the *tells*' archaeological record. The incorporation of a place required a process of transformation of the *quality* of the materiality of that place. This was achieved by compaction and levelling of the materials of the old inhabited surface, a process that certainly has had a significant ritual function.(Fig. 12)

Fig. 12: *Several layers of burned dwellings separated by levelling layers (Hârşova tell, 2005)*

Another type of technical ritual to be found in tell settlements is represented by the *chaînes-opératoires* of construction. Due to its predetermined structure by the laws of physics and the character of materials, a *chaîne-opératoire* can be perceived as a ritual operation (Gheorghiu 2011).

Since a ritual is perceived as a repetitive action, an established routine, or the result of a repetitive or structured behaviour (van Gennepp 1960 [1908]; Turner 1995), or of a stereotyped and redundant activity (Tambiah 1985), it is analogical with technology. From all the complexity characterising a ritual, only the quantifiable characteristics mentioned above can be identifiable in the prehistoric material culture.

Conclusions or Recycling as Ritual

The experiments with the construction and deconstruction of wattle-and-daub buildings, together with the data collected from ethnographic surveys, support the idea the processes of recycling identified in tell settlements were rites of passage, including a stage of utilisation, a liminal stage, and one of transformation and incorporation. This statement is valid at the same time for simple materials of building as well as for places. Perceived from this ritual perspective, a *tell* settlement is the result of a ritualized behaviour (Bell 1992) of recycling, including numerous stages of construction and deconstruction.

A first stage was represented by the founding operations of taking into possession of a chosen place, entailing the use of symbolic geometric patterns like the orthogonal frame (Gheorghiu 2009b), and the attraction of diverse materials from specific parts of the surrounding landscape, creating a sort of Mandala symbolism (Gheorghiu 2008).

All these operations, combined with that of recycling are to be found in every stage of dwelling in a *tell* settlement, and, since their structure was standardised and repetitive, they can be identified with ritual actions. A large part of the community participated to these ritual operations, because a *tell* settlement required a preconceived master plan and the rise of all the buildings at the same time, at least on the first level of dwelling.

Similar to the process of construction, the process of recycling can be perceived as a collective ritual, since it involved all the community when it returned cyclically to the old place of dwelling. The cyclical repetition of structured actions of utilisation and re-utilisation of the material of a single place, which is the result of a site catchment, can be seen as a ritual action of the recycling of a whole landscape, as well as one of ritual incorporation.

I will conclude stressing that a *tell* settlement seems to me the perfect example to demonstrate the importance of recycling in prehistory, and an efficient illustration of a syncretism between technology, symbolism, and ritual.

Acknowledgements

The experiments with prehistoric houses were financed by two CNCSIS research grants and partially by the Cucuteni Foundation for the 3rd millennium. The study of the rituality of tells was part of the Time Maps PN II IDEI research grant financed by UEFISCDI.

Bibliography

BAILEY, D. W.
   1997 Impermanence and flux in the landscape of early agricultural South Eastern Europe, pp. 41-58. In Chapman, J. and Dolukhanov, P. (eds.) *Landscapes in flux. Central and Eastern Europe in Antiquity*, Colloquia Pontica 3, Oxford: Oxbow Books.

BAILEY, D. W., ANDREESCU, R., HOWARD, A. J., MACKLIN, M. G., and MILLS, S.
   2000 Alluvial landscapes in the temperate Balkan Neolithic: transition to tells, *Antiquity* 76: 349-355.

BELL, C.
   1992 *Ritual theory, ritual practice*, New York, Oxford: Oxford University Press.

BRADLEY, R.
   2005 *Ritual and domestic life in prehistoric Europe*, London and New York: Routledge.

CANEVA, I.
   1999 Early Farmers on the Cilician Coast: Yumuktepe in the Seventh Millennium BC. pp. 105-114. In Özdoğan, M. and Başgelen, N. (eds.) *Neolithic in Turkey. The Cradle of Civilization. New Discoveries*, Istanbul: Arkeoloji ve Sanat Yayinlari.

CHAPMAN, J.
   1990 Social inequality on Bulgarian tells and the Varna problem. pp. 49-98. In Samson, R. (ed.) *The Social Archaeology of houses*. Edinburgh: Edinburgh University Press.
   1991a Social inequality on Bulgarian tells and the Varna problem. pp. 49-92. In Samson R. (ed.) *The Social Archaeology of houses*. Edinburgh: Edinburgh University Press.
   1991b The Early Balkan village. pp: 79-99. In Grøn, O., Engelstad, E., and Lindblom, I. (eds.) *Social space. Human social behaviour in dwelling and settlements*. Odense: Odense University Press.
   1999 Deliberate house-burning in the prehistory of Central and Eastern Europe, pp. 113-126. In Gustafsson, A. and Karlsson, H. (eds.) *Glyfer och arkeologiska rum - en vanbok till Jarl Nordbladh*, Gotarc Series A, Vol. 3. Göteborg: Institute of Archaeology, University of Göteborg.

COMȘA, E.
   1986 Santurile de aparare ale asezarilor neolitice de la Radovanu. *Cultură și Civilizație la Dunărea de Jos* 2: 61-67.
   1990 Complexul Neolitic de la Radovanu, Călărași. *Cultură și Civilizație la Dunărea de Jos* VIII.
   1997 Tipurile de asezari din epoca neolitica din Muntenia, *Cultură și civilizație la Dunarea de Jos* XV: 144-164.

CREANGĂ, E., CIOTOIU, I., GHEORGHIU, D., and NASH, G.
   2010 Vernacular architecture as model for today design, *International Journal of Design & Nature and Ecodynamics*: 157-172. WIT Press.

DUMITRESCU, Vl.
   1986 Stratigrafia așezării-tell de pe ostrovelul de la Căscioarele. *Cultură și Civilizație la Dunărea de Jos* 2: 73-81.

ERDOĞU, B.
   2005 *Prehistoric Settlements of Eastern Thrace. A Reconsideration*. British Archaeological Reports International Series 1424. Oxford: Archaeopress.

GHEORGHIU, D.
   2002 On Palisades, Houses, Vases and Miniatures: the Formative Processes and Metaphors of Chalcolithic Tells. pp. 93-117. In Gibson, A. (ed.) *Behind Wooden Walls: Neolithic Palisaded Enclosures in Europe*, BAR International Series 1013 Oxford: Archaeopress.
   2003 Water, tells and textures: A Multiscalar approach to Gumelnita hydrostrategies. pp. 39-56. In Gheorghiu, D. (ed.) *Chalcolithic and Early Bronze Age Hydrostrategies*, BAR International Series 1123 Oxford: Archaeopress.
   2005 *The Archaeology of Dwellings. Theory and Experiments*, Bucharest: Editura Universității București.
   2006a The Formation of Tells in the Lower Danube Wetland of Late Neolithic., *Journal of Wetland Archaeology* 6: 3-18.
   2006b Compactness and void: Addition and subtraction as fundamental operations in South East European Clay Cultures. pp. 151-161. In Frère-Sautot, M-C. (ed.) *Des Trous... structures en creux pre- et protohistoriques*, Actes du colloque international Archéologie et aménagement (Dijon et Baume-les-Messieurs, 24 - 26 mars 2006), Préhistoire 12, Montagnac: Ed. Monique Mergoil.
   2007a Material, spațiu, simbol. Note despre tehnologia de construcție și deconstrucție a locuințelor Chalcolitice din sud-estul Europei. *Anuarul Muzeului Etnografic al Transilvaniei*, pp. 36-379.
   2007b A Fire Cult in South European Chalcolithic Traditions? On the Relationship between Ritual Contexts and the Instrumentality of Fire. pp. 267-282. In Malone, C. and Barrowclough, D. (eds.) *Cult in Context*. Oxford: Oxbow Books.
   2008 Prehistoric Mandalas: The Semiosis of Landscape and the Emergence of Stratified Society in the South-Eastern European Chalcolithic. pp. 85-95. In Nash, G. and Children, G. (eds.) *The Archaeology of Semiotics and the Social Order of Things*, BAR International Series 1833. Oxford: Archaeopress.
   2009a The Lower Danube Chalcolithic megaron house with internal column: The technology of

building interpreted through experiments. pp. 1-10. In Ayan, X., Manana, P. and Blanco, R. (eds.) *Archeotecture: Second floor*, British Archaeological Reports, International Series 1971, Oxford: Archaeopress.

2009b The Symbolic construction and representation of the dwelt space in the Lower Danube Chalcolithic (6th millennium BC). pp. 113-118. In Djindjian, F. and Oosterbeek, L. (eds.) *Territories, travels and site locations. Symbolic Spaces in Prehistoric Art / Espaces symboliques dans l'art préhistorique. Territoires, déplacements et localisation des sites*, British Archaeological Reports, International Series 1999. Oxford: Archaeopress.

2011 Ritual chains, pp. 143-146. In Scarcella, S. (ed.) *Archaeological ceramics: A review of current research*, BAR International Series 2193. Oxford: Archaeopress.

2013 Space and Place as Artefact: On the Life and Death of Tell Settlements of the South Eastern Europe Chalcolithic. pp. 163-182. In Gheorghiu, D. And Nash, G. (eds.) *Place as Material Culture. Objects, Geographies and the Construction of Time*. Newcastle upon Tyne: Cambridge Scholars Publishing.

2014 Building, Burning, Digging and Imagining: Trying to Approach the Prehistoric Dwelling. Experiments Conducted by the National University of Arts in Romania, pp. 215-232. In Reeves Flores, J. and R.P. Paardekooper (eds.), *Experiments Past. Histories of Experimental Archaeology*. Leiden: Sidestone Press.

2016 Building and burning: The construction and combustion of Chalcolithic dwellings in the Lower Danube and Eastern Carpathian Areas fom the perspective of experimental archaeology. pp. 33-52. In: Nikolova L., Merlini, M., and Comşa, A. (eds.) *Western-Pontic Culture Ambience and Pattern. In Memory of Eugen Comşa*. Berlin and Boston: Walter de Gruyter GmbH.

GHINOIU, I. (ed.)
2003 *Atlasul etnografic român, Habitatul*, vol. 1, Bucharest: Editura Academiei Române.

HADJRI, K., OSMANI, M., BAICHE, B. and CHIFUNDA, C.
2007 Attitude towards earth building for Zambian housing provision. *Proceedings of the ICE institution of civil engineers, engineering sustainability* 160, issue ES3.

HAITA, C.
1997 Micromorphological study, *Cercetări arheologice* X: 85-92.
2000 Sedimentologie, *Cercetări arheologice* XI(I): 48-55.

HURCOMBE, L. M.
2007 *Archaeological Artefacts As Material Culture*. Oxon: Routledge.

KALICZ, N. and RACZKY, P.
1990 Das Spätneolithikum im Theißgebiet. Eine Übersicht zum heutigen Forschungsstand aufgrund der neuesten Ausgrabungen, pp. 11-33. In Meier-Arendt, W. (ed.) *Alltag und Religion. Jungsteinzeit in Ost-Ungarn*, Frankfurt on Main.

MARINESCU-BÎLCU, S., POPOVICI, D., TROHANI, G., and ANDREESCU, R.
1997 Archaeological researches at Bordusani - Popina (1993-1994). *Cercetări Arheologice* 10: 65-69.

Meşterul Manole (The Ballad of Master Manole)
1976 Translated by Dan Duţescu. Bucharest: Albatros Publishing House.

MINKE, G.
2006 *Building with earth. Design and technology for a sustainable architecture*, Basel, Berlin, Boston: Birkhäuser.

MINOIU, M. R., CIOBANEL, A. I., and BUDIS, M.
2005 Locuinta, In Ghinoiu, I. (ed.), *Habitatul*, Vol. 1, Oltenia. Bucharest: Editura Etnologica.

MORINTZ, S.
1962 Tipuri de aşezari şi sisteme de fortificaţie şi imprejmuiri in cultura Gumelniţa. *Studii şi Cercetari de Istorie Veche* 13 (2): 273-284.

MORINTZ, S. A.
2007 *Forme de habitat ale Eneoliticului final şi perioadei de tranziţie de la Dunărea de Jos*. Târgovişte: Cetatea de Scaun.

NEAGU, M.
1997 Comunităţile Bolintineanu in Câmpia Dunării, *Istros* VlII.

ÖZDOĞAN, M. and BAŞGELEN, N. (eds.)
1999 *Neolithic in Turkey. The Cradle of Civilization*. Istanbul: New Discoveries.

PERLÈS, C.
2001 *The Early Neolithic in Greece*, Cambridge: Cambridge University Press.

POPOVICI, D., RANDOIN, B., RAILLAND, Y., VOINEA, V., VLAD, F., BEM, C., and HAITA, G.
2000 Les recherches archéologiques du tell de Hârşova (dep. De Contantza) 1997-1998, *Cercetări Arheologice* XI(I): 13-34.

REDDY, B. V. V.
2007 Indian standard code of practice for manufacture and use of stabilised mud blocks for masonry. *International Symposium on Earthen Structures, Indian Institute of Science, Bangalore, 22-24 August*. Interline Publishing, India.

SHERRATT, A. G.
1983 The Eneolithic Period in Bulgaria in its European Context, pp. 188-198. In Poulter, A. G. (ed.), *Ancient Bulgaria. Papers presented to the International Symposium on the Ancient History and Archaeology of Bulgaria, University of Nottingham, 1981*, Nottingham.

STEVANOVIČ, M.
1997 The Age of clay. The Social dynamics of house construction, *Journal of Anthropological Archaeology* 16: 334-395.
2002 Burned Houses in the Neolithic of Southeastern Europe. pp. 55-62. In Gheorghiu, D. (ed.) *Fire in Archaeology*, BAR International Series 1098. Oxford: Archaeopress.

ŞTEFAN, C. E.
2010 *Settlement types and enclosures in the Gumelniţa culture*, Târgovişte: Cetatea de Scaun.

TAMBIAH, S.
1985 A Performative Approach to Ritual. pp. 122-166. In Tambiah, S. (ed.) *Culture, Thought and Social Action*. Cambridge, MA: Harvard University Press.

TODOROVA, H.
1982 *Kupferzeitliche Siedlungen in Nordostbulgarien*, Muenchen: C.H.Beck.
1978 *The Eneolithic in Bulgaria*, Oxford: British Archaeological Reports.

TRINGHAM, R.
1992 Households with faces: The Challenge of Gender in Prehistoric Architectural Remains, pp. 93-131. In: Gero, J. and Conkey, M. (eds.), *Engendering Archaeology. Women in Prehistory*. Oxford and Cambridge: Blackwell.

TRINGHAM, R., and KRSTIČ, D.
1990 Conclusion. Selevač in the wider context of European prehistory. pp. 567-617. In: R. Tringham si D. Krstic (ed.), *Selevač. A Neolithic village in Yugoslavia*, Monumenta Archaeologica 15, Los Angeles: University of California Press.

TURNER, V. W.
1995 *The Ritual Process: Structure and Anti-Structure*. New York: Aldine de Gruyter.

URSULESCU N., TENCARIU, F-A. and BODI, G.
2002 Despre problema construirii locuinţelor cucuteniene, *Carpica* XXXII: 5-18.

URSULESCU, N., TENCARIU, F-A., and MERLAN, V.
2002 Noi date privind sistemul de fixare a pereţilor in cultura Precucuteni, *Carpica* XXXI: 13-21.

VAN GENNEPP, A.
1960 (1908) *The rites of passage*, London: Routledge & Kegan Paul.

VARGAS, J., BLONDET, M., GINOCCHIO, F. and VILLA-GARCIA, G.
2005 35 years of research on SismoAdobe. *International Seminar on Architecture, Construction and Conservation of earthen buildings in seismic areas. SismoAdobe 2005*. Lima, Peru: PUCP.

WHITTLE, A.
1996 *Europe in the Neolithic. The creation of new worlds*. Cambridge: Cambridge University Press.

Copper and Bronzes:
The Birth of Complete Recycling in The Bronze Age

Davide Delfino

Abstract

The Archaeology of production has explored the theme of the life cycle of archaeological materials, highlighting some stages after use, of which recycling is a part: metals, particularly ancient copper alloys, are of particular interest in this phase. The discovery of copper metallurgy revolutionized material production and other aspects of social life. The time and expertise required for its reduction from ore along with the possibility of being recast, led to the custom, and somehow the convenience, of recycling probably by the Chalcolithic. In the Bronze Age the art of metal alloying assumed even greater value: greater knowledge and rare metals, such as tin, were needed in the preparation of the alloy, whilst metal recycling became even more entrenched. Archaeology, and above all the above-mentioned "Archaeology of production" can rely on on several types of evidence to reconstruct and understand the techniques, dynamics and intentionality of the recycling of a precious material, which were loaded with symbolism by the human community in the "Metal Ages": founder's hoards, production sites, archaeometallurgical data. The study of recycling of bronze can help in the understanding of the social and economic dynamics and resource management of the communities of the European Bronze Age.

Introduction

The theme of metal recycling in early history is of primary importance in the understanding of the management of a basic material for the societies of the "Metal Ages", but has some other specific difficulties:

- the vast European panorama, not only from a geographical point of view, but also chronological and cultural;
- the difficulty of putting together data of different types, archaeological and archaeometrical;
- having to work on material evidence that until recently were held in poor regard by archaeologists and do not always come to us with sufficient documentation;
- the problem that, as the material is completely recyclable, scrap bronze remains are under represented with respect to the amount really used in the recycling.

There have been few specific studies on the subject of the recycling of metals in Prehistory and Proto-history. One of the few is the work of Needham (1998), where in an analysis of models of "metallurgical" regions, particular attention is paid to analysis of the weight that recycling has on the type and method of metallurgical production (Needham 1998: 290-291); this interpretative model recognizes three types of "metallurgical region", which are:

1. *recipient only*: a region that receives finished products, but does not have the expertise to produce metal artefacts;
2. *cast-once-only*: a region where the technical expertise to produce metal artefacts is present, but where imported objects are not recycled more than once;
3. *recycling*: a region where the metal of the same objects is reused several times, which is where recycling is a core activity for metallurgical production.

Each model of a region depends on the characteristics of natural resources (metal minerals) and metallurgical working capacity, factors considered by Needham, but depends also on other important factors, such as geographical position and social structure. In truth this system of three models of metallurgical region appears somewhat "static" and "mainly theoretical", a fault that is very common to many interpretative systematic "closed" models, applied to Prehistory and Protohistory. It is said for example that the recycled items are always imported (*cast-once-only* model region). Also the fact that the recycling of an object in a region takes place once, twice, or even three times is completely subjective: archaeometrical analysis cannot understand with certainty how many times a metal was re-melted and recycled.

Whatever the case, the "recycling factor" is important in the understanding of technological capabilities, metallurgical economy and circulation of metal artefacts in one or more regions.

Metals: properties interesting for recycling

Among all the materials offered by nature and transformed by man according to his needs, metals are those that can be recycled more quickly, using simple technical processes. They do not need complex processes of transformation and, after being recovered, can almost immediately be used for new artefacts. However, after each successful they still require a few additional treatments: filtering waste slag formed and the addition of small amounts of additives to make good the combination for recycling.

What makes metals, and especially those used since recent Prehistory, such as copper or gold, is the crystal structure and physical-chemical properties that characterize them.

The chemical properties are related to atomic structure, while the physical and mechanical properties are related to the crystal structure. Solid-state stiffness of the crystal structure is guaranteed by the enormous intensity with which the inter-atomic forces inhibit the movement of individual atoms.

In metals, the crystal structure is determined by intermolecular forces. If they are subject to a certain temperature, they break (Giardino 2010: 23-24)

In particular, the physical and chemical properties of metals that make them an excellent recyclable material are twofold:

1. Metals have a very specific melting temperature: the melting point is reached because it achieves a sufficient amount of thermal energy to break the intermolecular forces (Giardino 2010: 16).
2. Metals differ from other materials, and can easily be recycled, specifically because the fusion of the crystal lattice bonds is broken, transforming the solid to liquid. On cooling, the atoms of the metals have the specific property of tending to draw closer and then they quickly solidify the molten metal and recompose the crystal structure (Giardino 2010: 26).

Other mechanical properties of metals make them an easy material to work with features that allow you to develop an advanced manufacturing technology: a good relationship between hardness and toughness; ductility.

These features, that are not found in either organic (bone, leather) or in lithic material, made metal a revolutionary material and essential to the human communities of recent Prehistory. Even after this, as we shall see later, metal was often recycled and not just because it was recyclable.

Metals: diffusion of ore used in Bronze Age in Europe

As far as recycling is concerned, the presence of natural resources in a territory can have a double meaning; when an area has an availability of, for example, copper ore, it is almost certain that the human community that lived there had a sufficient supply of raw material and should not be compelled to make large copper-bronze objects through recycling, but it is equally true that even in these circumstances there would be a certain level of recycling to optimize the exploitation of natural resources.

The metals are present in very weak concentrations in the Earth's crust. Average values for one of the metals of interest in prehistory and proto-history – copper, are 0.01% (Bartocci and Marianeschi 1960: 22-23). All metals would be very rare if there were concentrations of up to 10,000 times higher than the average values, in the vicinity of outcrops of veins due to local enrichment processes. (Bartocci and Marianeschi 1960: 22-23).

The concentration of ores exploited in prehistory and proto-history shows that copper is widespread in non-solid, but significant sources in different European regions, while tin is much less common (only in rich deposits in Cornwall, Britain, Galicia and Extremadura, and in poor deposits in Sardinia, Tuscany, Devonshire, Limousin, Charente, Creuse, Haute Vienne, Bourbounnaise, Erzgebire and Murcia) (Giardino 2010: 133-139).

The Archaeology of Production

The "Archaeology of Production" is crucial in studies of the recycling of materials It opens new horizons, both potentially in method and documentation. This technological approach pays attention to archaeological evidence that is less spectacular, such as production facilities, waste products, and traces of mineral extraction or refining. An excellent case study on the subject is that of T. Mannoni and E. Giannichedda (1996), probably the only complete work on the subject published so far. In "The Archaeology of Production", the production cycles are very important: in particular, in the final stages of each cycle, that is of use, breaking and recycling. This is especially so in the cycle of the metal recycling, which allows the craftsman to completely reshape the raw material recovered and the archaeologist to find substantial traces of this phase on the site of production/recycling. Objects that are used in recycling and can be archaeological evidence of this are:

1. Objects that are perfect products but become unusable with use; those entering and leaving the cycle of use; for example an axe with a chipped blade.
2. Objects that did not achieve the desired characteristics by accident during processing and are defined as "defects" (*ibid*: 193); those never entering in the cycle of use and just entering non-use cycle; for example a poorly cast axe.

3. Artefacts that are inevitable as a result of the production cycle that leads to the final product, that "waste" (*ibid.*); those not thought to enter in the cycle of use and therefore not entered in the non-use cycle; for example casting droplets.

The "waste products" are not very relevant in the study of recycling in metallurgy, as here the possibility of their total recycling is higher (for example, in respect of the lithic industry) and are generally not preserved in archaeological contexts, except for special cases. In the "defects" there is a risk of an overrepresentation of more complex products that require above all a more elaborate casting mould: they certainly produce more defective parts than others, which are under-represented. Interpreting and not only recognizing, the finds from a production cycle, or use cycle, entails an understanding of what the ancient craftsman attributed to them: if they are recycled items, even if it is "waste" in technical jargon, one will need to speak instead of "recoveries". Privileged contexts for investigation in recycling are production sites, where some "recoveries" are at times saved from melting and recycling storage sites, where the "recoveries" were collected. The production sites that potentially have a lot of information on economic cycles were linked to those, which may fortuitously, help in the understanding of the link between the collection of scrap for recycling and the economic circuit.

Anthropological Questions

Cultural anthropology attaches great importance to technological choice: it emphasizes the ability and capacity of human society to choose between a number of alternatives; these are related to environmental factors, political factors and ideological factors (Lemonnier 1993). In addition to these factors, the more complex the production cycle, the longer the learning phase and hence the greater the desire to control the spread of knowledge (Conati Barbaro 2005: 174). This is particularly true for the cycle of metals, which requires complex steps, skills and knowledge.

Archaeological Evidence

The European Bronze Age is characterized by various macro-regional areas, which, by their very large metallurgical production and affinity of artefacts, are known as "Metallurgical circles" (Carancini, Peroni 1997). It will now be necessary to examine some of these macro-regions, according to their activities in the recycling of metals, particularly bronze. To achieve the most comprehensive and complete overview, consideration will be undertaken of areas near some of the major copper sources that were subject to massive metallurgical production in Europe (in this case some metallurgical circles in Southern Europe), as well as "marginal" areas.

The archaeological evidence for the recycling of metals in the Bronze Age is mostly made up of founder hoards. Of all the forms of metal storage (ritual offerings, hoards or founder's deposits), those of the founder identify themselves by three factors: scrap objects, fragments of ingots of copper or bronze and place of deposition (in a ceramic pot, or in niches in bedrock or in pits). Why preference be given to founder's hoards for the analysis of recycling? Certainly there are contexts that have left more traces of this activity, which is an integral part of the copper/bronze cycle. Some aspects of the interpretation and study of the deposition of metalwork must be highlighted, as has been done in important studies and contributions on this argument (Bradley 1988; Delibes de Castro 1997; Harding 2000: 323, 352-368; Vilaça 2006). Before observing some manifestations of scrap hoards, it is appropriate to point out some of the problems and issues. The concept now widely accepted is that hoards are a dynamic and complex phenomenon, as a result of multiple events of deposition and collection. This is particularly true of the hoards with scrap objects for recasting. We must also take into account the fact that many hoards were found in the past (between the nineteenth and early twentieth centuries), which means that their original composition is unknown: some examples are emblematic from this point of view, e.g. Portugal (Vilaça 2006: 29-37); and Italy (comparing the number of artefacts in depots reported on discovery and that of the objects reported in recent publications).

The Scrap Hoards outside the Settlement

The Alps and Northern Italy

There are several copper deposits in the Alpine region (Figs. 1A and 2). The known sites in the Western Alps are as follows: Alagna, Brosso (Cierny 1997: 77), Murialdo, Bormida, Biestro (Pipino 2008: 51), the Tournanche valley (Valle d' Aosta), Alpi di Pinerolo, Sesia valley, Chiusella valley (Piedmont), Venerocollina valley (Lombardy) and Agordo (Venetian) (Giardino 2010: 116) in Italy, and Dôme de Barrot, Le Cerisier, Dalvis, Le Grange du Colonel, Giraud and the mine of St. Verán with clear evidence of prehistoric works (Cierny 1997: 76) in France. The known sites in the Central Alps are those of Predoi e Valsugana, in Italy (Cierny 1997: 77) and the Mitterberg, Pinzgau, Schwaz, Brixlegg, Falkenstein, Kitzbuhel and Kallwang, in Austria (Cierny 1997: 76), of which the Mitterberg (Middle Bronze Age), Kitzbuhel and Brixlegg (Final Bronze Age) were certainly used in prehistory (Cierny 1997: 78).

Known traces of intense copper mineral extraction activity in the Southern Alps indicate that the periods between the Chalcolithic and the end of Early Bronze Age and between the Late Bronze and the Final Bronze Age were periods of greater intensity of exploitation

Fig. 1: *Investigated areas; Alpine region (A) and Atlantic Iberian Peninsula (B)*

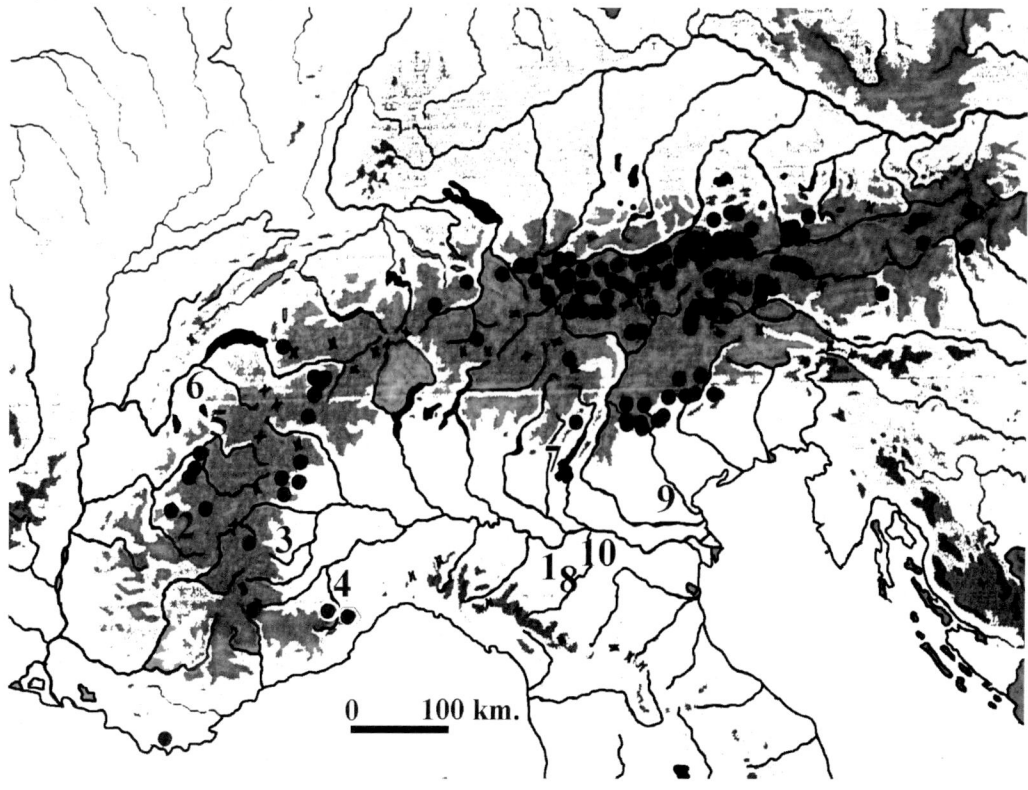

Fig. 2: *Alpine region: the copper resources (black circles) and cited sites (1. Baragalla; 2. Casse Rousse; 3. Pinerolo; 4. Bric della Sorte; 5. St. Pierre d' Albigny; 6. Meytet; 7.Lugana Vecchia; 8. Castellarano; 9. Frattesina di Fratta Polesine; 10. S. Francesco di Bologna) (Elaborate from Cierny 1997: 77)*

(Marzatico 1997: 572). This is particularly true of the Trentino Alps, where a large number of copper ore reduction sites have been found. It seems that the copper in the valley was exploited in the first period, whilst those at higher altitudes were exploited in the second period (Cierny *et al.* 1998: 26-32)

Examining the evidence in chronological order, the founder's hoard of *Baragalla (Reggio Emilia)* (Early Bronze Age II- 1750/ 1600 B.P.) is certainly the oldest evidence of bronze recycling in northern Italy (De Marinis 2006a: 1295) and probably in this macro region. Scrap objects are documented with small ingots of plano-convex shape here for the first time. The movement of scrap objects typologically typical of the Franco-Swiss area is particularly clear on this site. The hoard was made up of bronzes just over 3 kg in weight (9 axes and 21 small raw copper ingots integers) (De Marinis 1997: 306-307). This phenomenon has been interpreted as an index of the large-scale movement of scrap objects by De Marinis (2006a: 1295). However, it can also be interpreted as the movement of finished products, used in places distant from the production area, and then recycled.

The Late Bronze Age (1325-1175 BC) is the period, in which the number of founder's hoards with a presence of the largest raw copper ingots began to increase. This includes metals groups that are not real hoards, but more clusters of scrap metal and raw copper ingots found in the vicinity of metallurgical workshops.

In general, there was a proliferation of the production of bronze in the Late Bronze Age in Italy, which lasted until the Early Iron Age. It would appear that the main indicator for this theory is the proliferation of metal hoards, especially scrap hoards (Giardino 2005: 491)

During the Late Bronze Age, many hoards of scrap bronze objects, together with fragments of copper ingots start to appear in the Western Alps and adjacent areas, especially in the French Alps (Provence-Alpes-Maritimes-Haute Savoye-Savoye-Isère) and the Italian Western Alps. Eleven scrap and ingot hoards are known, particularly the sites of Lullin-Couvaloux, Sion, La Poype Vaugris and Casse Rousse (France) and Pinerolo, Pietramazzi, Coniolo, Bric del Ciaz and Bric della Sorte (Italy) (Barge 2004; Bocquet, Lebascle 1983; Courtois 1960; Del Lucchese, Delfino 2008; Doro 1975; Garcia 2003; Gambari and Venturino Gambari 1994; Mordant 2003). Three of these sites are of particular interest:

La Croupe de Casse Rousse (Briançon): 83 pieces of bronze, together with fragments of pottery, slag and charred vegetal remains were found on top of a steep cliff at a

Fig. 3: *Bric della Sorte hoard (Savona-Liguria) (Image of Soprintendenza per i Beni Archeologici della Liguria).*

height of 2.070 m above sea level (Rossi and Gattiglia 1998). Pieces of scrap bronze and raw copper ingots are clear indicators of recycling activity, but in this case as at the metal workshop of Lugana Vecchia, it is not possible consider it a true hoard.

Pinerolo (Torino): a group of bronze objects that can be called a classic "founder's hoard" was discovered at the mouth of the Pelice and Chisone valleys on the Italian side of the Alps. It comprised axes, armring fragments, sickles and raw copper ingots (Doro 1975). It is clear evidence of the use of recycled items that were typical of the Western Alps and circulating between the two sides of the Alps: the fragments of "La Poype" type armrings in the hoard, that are also found in many hoards in the French Alps, are an example.

Bric della Sorte (Savona) (Fig. 3): in the Ligurian Alps, a hoard composed of small fragments was discovered in a hollow in the rock on a mountain in a valley that links the Po Valley and the northern Tyrrhenian Sea. It comprised 21 finished objects and 50 raw copper ingots (Del Lucchese, Delfino 2008). The few artefacts that could be recognized typologically date the deposit to the Late Bronze Age. It also contains metal objects typical of the Western Alpine Metallurgical Area: a fragment of a "La Poype" type armring, in particular, is a clear indicator of this.

The Final Bronze Age (1175-750 B.C.) did not only see a continuous increase in the hoard containing items to be recast, but above all an increase in their size, in terms of number and weight of objects. Some examples are important in the understanding of the dimension of the phenomenon of scrap bronze collection during this period; amongst others it sees the disappearance of the one pre-urban civilisation in the area (the Terramare), the onset of a period of crisis (with some exceptions in the Western and Eastern Alps) and the birth of the cultural substratum that will lead to the Villanova Culture. Some founder's hoards are still found in this period in the Western Alps:

Saint-Pierre d'Albigny (Albertville-Savoie) (Bocquet and Lebascles 1983: 48-51): unfortunately, some of the original parts are missing. The hoard was found in a mountainous area, in one of the main valleys communicating between Italy and France. The hoard was composed of a single damaged axe, two "pick-ingot" fragments, two fragments of sickles and an armring fragment. The type of objects date the hoard to the 10th-9th century BC. They are objects from both Western Alpine and Northern Italian tradition, showing possible trafficking in scrap between the two regions.

Meytet (Annecy-Haute Savoie) (Bocquet and Lebascles 1983: 46-48): found in a pottery vessel. It contained two axes, fragments of five axes, three fragments of swords, four sickles, a sickle fragment, three armring fragments, a pin and some pin fragments. The artefact typology dates the hoard to the 11th-10th century BC, with objects of regional manufacture.

Atlantic Iberian Peninsula

The Atlantic facade of the Iberian Peninsula (Figs. 1B and 4) is relatively rich in copper deposits, particularly in Huelva (Southwest Spain) and Asturias (Northwest Spain). However good deposits are also found in Portugal, in the Algarve, Baixo Alentejo, Estremadura, Beira Alta (Giardino 2010: 115-116) and Beira Baixa (Cardoso and Vilaça 2008: 44). There are also rich deposits of tin, in the Spanish Extremadura, Leon, Galicia (Spain), Beira Alta, Minho and Alto Douro (Portugal) (Cardoso and Vilaça 2008).

The Atlantic region of the Iberian Peninsula does not seem to have many hoards that have the characteristics of smelter sites for accumulation of recovered material (containing scrap and raw copper ingots), in proportion to the general number of bronze hoards. It has been shown that smelter and scrap hoards in Italy also appear in the Early and Middle Bronze Age, although in rare cases, only to greatly increase in the Late Bronze Age, whilst in the Atlantic part of the Iberian Peninsula this material appears exclusively in the Final Bronze Age (Brandherm 2007: 179).

In the general framework of the Final Bronze Age society, mention should be made of the analysis made by Senna-Martinez (2007: 266, 270-274). It appears that a system of population in small settlements with hierarchy between them existed in central Portugal (but also the Spanish Extremadura). There are some examples, which may be considered in this context:

The hoard of *Porto do Concelho (Mação)* (Jallay, 1944) (Fig. 5): this is a classic founder's hoard, composed of both scrap and complete objects from different periods of the Bronze Age (typologically between the Middle Bronze Age and the end of the Final Bronze Age). There are 39 objects in total: 18 small rings, some of which are very fragmented, three axes, including one broken example, 15 weapons, comprising swords, daggers and spearheads, including 11 broken examples, and, finally, a fragment of a fibula. R. Vilaça (2006: 84) has rightly pointed out the fact that the hoard also contains small fragments of very common objects (rings), suggesting that everything, even the most insignificant fragments, were recycled.

The hoard of *Quinta do Ervedal (Fundão)* (Coffyn 1985) consisting of 12 axes, including 9 in poor condition and three recently cast, 2 probable sword fragments, 4 fragments of unidentified objects and a small raw copper ingot, one of six. All the objects date to the Final Bronze Age, according to typological analysis. We are dealing with a smelter deposit, with clear evidence of recycling, but also of objects that were the result of this action (the axes with casting flash).

Attention must also be drawn to two other hoards in central Portugal:

Casal dos Fieis de Deus (Bombarral) (Vilaça 2006: 40-41): consists of 12 items: two fragments of swords, a spearhead fragment, an axe fragment, six massive armrings (one fragmented), and a dagger. There were no raw copper ingots. Based on the presence of swords of the "carp's tongue" type and a "gun" type axe the hoard can be dated to the end of the Late Bronze Age (9th-8th centuries BC).

Vila Cova de Perrinho (Vale de Cambra) (Vilaça 2006: 67): found in a pottery vessel, consists of two "ring" type axes (one fragmentary), a deformed dagger, a sword fragment, five fragments of unidentifiable bronze objects and 4 chisels.

The famous hoard of *Ria de Huelva* can be included in the various hoards that are composed in part or entirely of scrap items. An examination of its composition reveals that almost half of the objects are damaged. This hoard, however, reveals some other unknown factors at the interpretive level. Is it a votive deposit in water (*gewasserfunde*)? Or is it the load of a single shipwreck

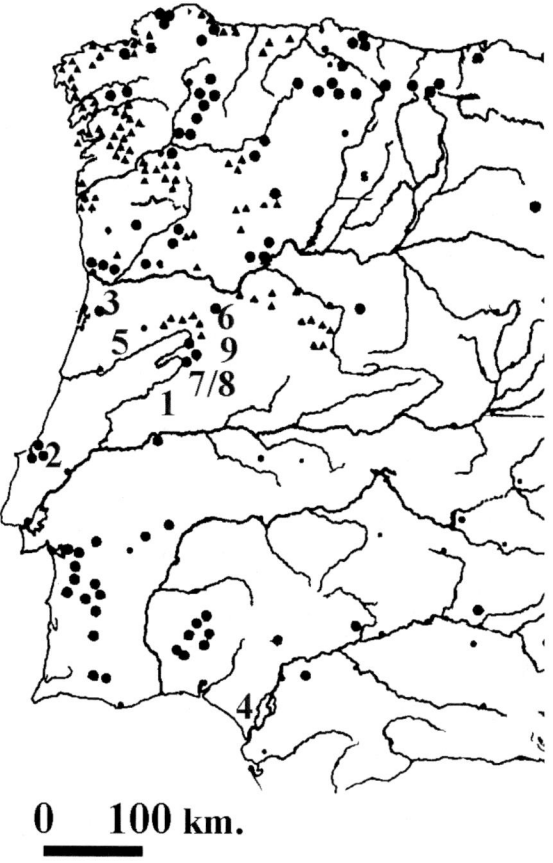

Fig. 4: *Atlantic Iberian Peninsula: the copper (circles) and tin (triangles) resources and citied sites (1. Quinta do Ervedal; 2. Casal dos Fieis de Deus; 3. Vila Cobva de Perrinho; 4. Ria Huelva; 5. Castro de Nossa Senhora de Guia; 6. Castelejo; 7/8. Alegrios/ Moreirinha; 9. Monte do Frade) (Elaborated from Cardoso, Vilaça 2008: 44).*

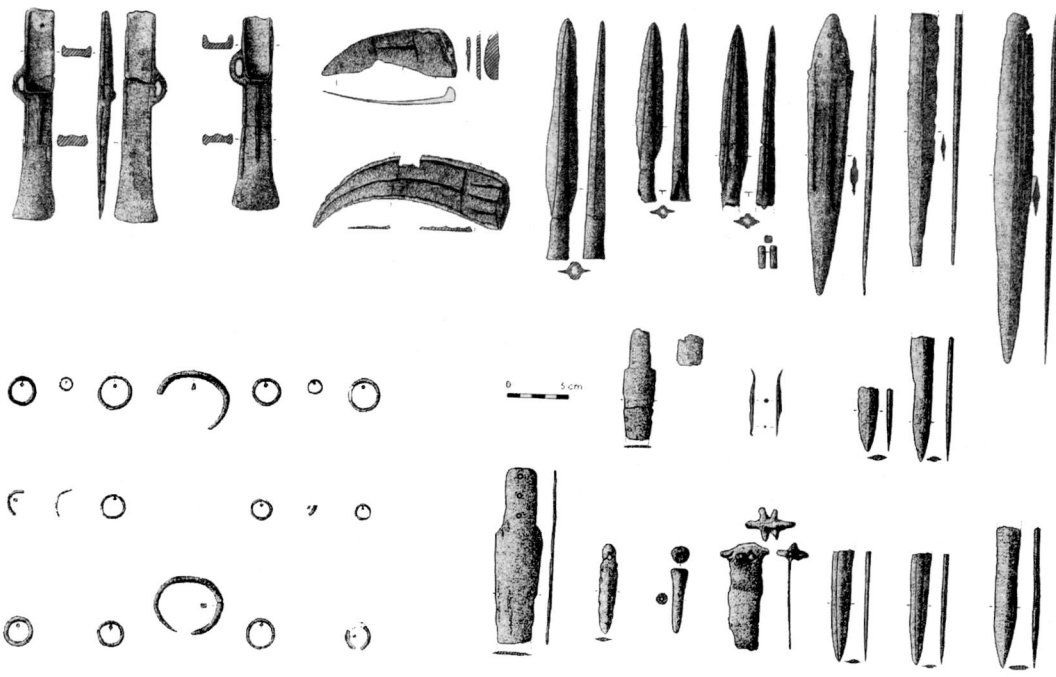

Fig. 5: *Porto do Concelho hoard (Mação- Pinhal Interior).*

or several shipwrecks of different periods? The first hypothesis, as a series of subsequent votives offerings is somewhat problematic: if one considers artefacts such as swords, it is baffling that this votive deposit contains pieces of objects showing signs of intentional breakage (blade broken in half), next to specimens with signs of damage that is probably not intentional (especially given the proximity to the hilt). However, it is certain, as far as this research is concerned, that it also comprised scrap objects in water, especially swords and spearheads. So it is not a typical founder's hoard. If one accepts the hypothesis that is the cargo of a ship, then one is confronted with evidence of trade by sea, or river routes, of scrap, possibly from outside the wider region. According to the lead isotope analysis that was carried out recently, it seems that in general the copper of the bronze in the hoard is not from the Pyritic Band (between Algarve and Huelva), but from mineralization that is further removed from Ria Huelva, particularly the scrap from the geological area of Ossa Morena and Sierra Morena (Montero Ruiz *et al.* 2007: 196-198, 203, 206, 208).

"Marginal" metallurgical areas

Three founder's hoards and several small local copper sources are known in Calabria in southern Italy and in the centre of the Mediterranean area One of the three hoards is from an unidentified site in the province of Crotone (Marino and Pizzitutti 2008). It contained 48 items, including 42 scrap artefacts and 6 raw copper ingots, with a total weight of 5 kg. The majority of items are scrap axes with weapons in the minority. This hoard, dated to the Late Bronze Age on basis of artefact typology, indicates the practice of recycling and deliberate fragmentation of objects. This demonstrates how the recycling aspect was not secondary in the economy (Marino and Pizzitutti 2008: 331).

Scrap Hoards in Settlements

Another type of founder's hoard is found not in isolation outside inhabited areas, even in inaccessible areas and in hidden areas, but in settlements, probably as part of metallurgical workshops. It is important to understand the dynamics of different types of scrap collection and storage, when considering the difference in context between the hoards and the data relating to the second group.

Some evidence for the recycling of metals appears at the end of the Middle Bronze Age (14th century BC), in the pile dwelling settlement areas in the eastern central part of the Po valley. One such example is the hoard in the pile dwelling settlement at *Lugana Vecchia* (*Brescia*) (Fig. 6), which contained scrap bronze objects, resulting from unsuccessful casting, slag, raw copper and tin ingots (De Marinis 2006a: 1295-1298).

Another example, dating to the Late Bronze Age in the Terramare area, is the "hoard" from *Castellarano* (*Reggio Emilia*). This is a collection of scrap consisting of fragments of axes, sickles, swords and pieces of "plano-convex" type copper ingots (De Marinis 2006a: 1299). In this case, as Raffaele Carlo de Marinis has pointed out (De Marinis 2006a: 1300), the ingots appear to be almost pure copper, which means that the scrap from bronze objects was not cast into new ingots, but was loaded directly into crucibles to produce new artefacts.

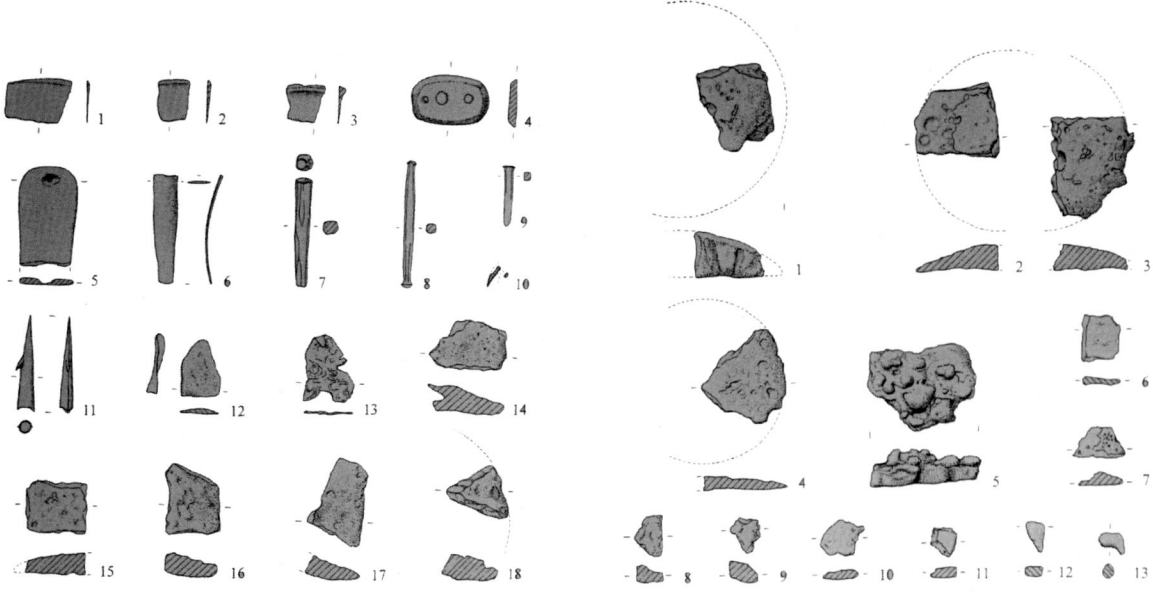

Fig. 6: *Lugana Vecchia scrap in settlement (Brescia- Lombardia) (Elaborate from De Marinis 2006a: 1298,1300).*

The settlement of *Frattesina di Fratta Polesine (Padova-Veneto)* is located in North-eastern Italy (Fig. 7). From the Late Bronze Age to the Final Bronze Age it was an important centre for trade between northern Europe and the Mediterranean, especially the trade and processing of Baltic amber. Some metallurgical workshops were found in the settlement. Four of theses have produced evidence of the recycling of scrap objects. One of the four, like the other three, contained 72 scrap objects, a copper ingot of "pickaxe" type, and a lead ingot. In general, axes and weapons are in the minority. The total weight of scrap objects and ingots is 6.58 kg., while the typology of the objects suggests a date in the Late Bronze Age II (10th-8th centuries BC) (Salzani 2003). Another one of these four metallurgical workshops consisted of 51 scrap objects and 21 pieces of raw copper ingots and ingots of 'pickaxe' type. There are no weapon fragments. This group is dated to the Late Bronze Age II as are the other three workshops (Salzani 2000). The presence of certain types suggests an extensive network for the movement of metals, including scrap items and ingots of impure copper, known as the 'pickaxe' type, in one of the largest storage centres in the Frattesina settlement (Salzani 2000: 46).

An even more important hoard for northern Italy, during the proto-historic period, in terms of number and total weight of objects present, is that of *Bologna- S. Francesco* (Antonacci Sampaolo et al. 1992). It was found in the historic centre of Bologna in 1877 and contained 14,841 objects, both scrap and complete, which weighed about 1,500 kg. It was found in a large vessel inside a hut. It comprises 12,185 items of various scrap objects, 1,050 pieces of waste metal and 1,606 raw copper ingots. The typology of the artefacts is very varied and suggests a date range for the hoard, the 'closest' being dated to the 10th-7th centuries BC. The large amount of data available here would make it possible to undertake many analyses and decades of study on this important discovery alone. It is possible to conclude that the scrap was further broken to better enable re-melting and that unfinished objects and tin (in ingot) indicate the main occupation in this workshop in the recycling of bronze (Antonacci Sampaolo et al. 1992: 177).

The Atlantic Iberian Peninsula

The hoard of *Castro da Senhora da Guia (Baiões)* (Senna Martinez 2007: 269; Kalb 1978) should rather be considered as a metallurgical atelier. It comprises various fragments of casting moulds and bronze that were discovered in a settlement. Some of these can be recognised as fragments of plates, which are clearly to scrap for recasting. This is a good example of a metallurgical workshop in a settlement, which also worked with the recycling of scrap bronzes. This occurred in a cultural context, which does not appear to attest a large scale movement of metal objects, but only movement at a local level (Senna Martinez 2007: 268, 274).

Some groups of objects of bronze were found in four settlements in the Beira Interior region (Portugal), which dating to Atlantic Final Bronze Age II. These are either scrap, had just come out of the casting mould and had casting defects, or were fragments of raw copper ingots (Vilaça 1997). The finds are as follows: two fragments of raw copper ingot, one sickle, one bar, in the settlement at *Castelejo* (Sabugal); a fibula, two rings, eight bars, tree daggers, one pin and three raw copper ingots in the settlement at *Alegrios* (Idanha-a-

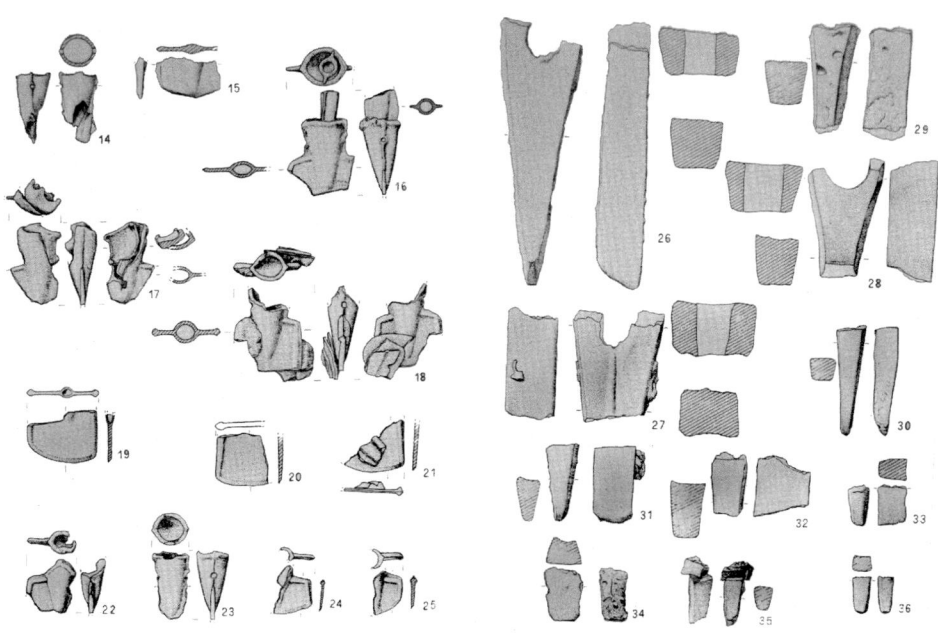

Fig. 7: *Frattesina di Fratta Polesine scrap in metallurgical workshop No. 4 (Padova-Veneto).*

Nova); two daggers, one razor, two bars and two arrow heads in the settlement at *Monte do Frade* (Penamacor); six daggers, twenty bars, one raw copper ingot, twelve rings, two buttons, one unidentified artefact, one fibula in the settlement at *Moreirinha* (Idanha-a-Nova).

The metal artefacts found in these settlements are considered to be objects that were poorly cast and were put aside. This can be considered in the context of the work of R. Vilaça (1997: 124) that compared the life of bronzes to that of human beings, in which he states that the object 'died in a growth phase'. However his assertion that they 'were abandoned' is questionable whether (Vilaça 1997). In fact, from the description of the context of the discovery (with traces of structures and combustion), there is no reason not to think of storage as a function of recycling in the same structures.

'Marginal' metallurgical areas

There are few copper local resources, but there are metallurgical workshops in the area of the Italian peninsular, which was occupied by the Apennine Culture in the 13th century BC. There are, for example, two metal workshops with scrap objects and a raw copper ingot in the settlement at Moscosi di Cingoli. An important feature of these finds is that all the scrap has a high percentage of tin in the bronze alloy (De Marinis 2005).

Other Archaeological and Archaeometrical Data

In order to conclude without ignoring important data from outside the above areas, one must also take under consideration data from other important geographical areas, such as the British Isles (Mutti 1993: 167), where smith's hoards with scrap objects increase from the Middle Bronze Age to Late Bronze Age, as in the Alpine/Northern Italy region and the Atlantic area of the Iberian Peninsula. Another important fact that must be considered concerns the "plano-convex" type copper ingot. These are almost always present in the smith's hoards, along with scrap for recycling. The classic type, 15/20-19/23 cm. in diameter and 1/9.5-7/10 kg. in weight, is in constant use in Mediterranean and Continental Europe from the Late Bronze Age (De Marinis 2006a: 1300) to the Iron Age (Domergue 2004). In almost all cases, the ingots of this type are more or less pure copper, which means that scrap bronze objects were recycled by adding a little pure copper (De Marinis 2006a: 1300) and perhaps a small amount of tin. This fits together chronologically throughout the entire European area with the increasing number of hoards with scrap to recasting and "plano-convex" ingot types. In the face of this evidence this leads to the following observation: the increase of recycling (caused by low availability of raw material?) and of production of "new" copper appear to be synchronic. This should lead to some considerations that will be proposed later. With more specific regard to the problem of the metallurgical technology related to recycling, there is evidence from many of the Later and Final Bronze Age contexts, in, for example, Italy, that the bronzes are highly uneven in composition. This is probably due to the massive use of recycling (Giardino 2005: 491), which generates a considerable mixing of secondary elements and primary elements. Thus a problem that arises with regard to this phenomenon that can lead to difficulties in the control of the bronze alloy obtained after the recycling cast is the quality of the final alloy. Whereas the properties conferred in the percentage of tin in bronze alloy, the problem for the smelter recycling the scrap (of different alloy composition) and obtaining a new alloy, was to ensure for the potential customer a good material that can be processed (Fasnacht 1997: 84).

Looking at the archaeometallurgical data from a "marginal" region for metallurgy (Abruzzo in peninsular Italy) (Bietti *et al.* 2003), one can observe that there was a good production of bronzes with a good percentage of tin from the Late Bronze Age. Many objects can be observed with low iron content: in most cases this is the balance of slag from the smelting process. When it is found in low percentages it can be interpreted as an index of recovery performed to obtain the object in question (Bietti *et al.*2003: 411-425)

Using data from one of the most important finds for the study of trade in the Bronze Age, the Ulu Burun ship wreck (13th century BC), it is possible to observe that raw copper ingots (not of the "oxide" type, but "plano-convex" type) have a poor quality of purity: this leads to the hypothesis that copper would still be subjected to rough additions on use (Lo Schiavo 2006: 1321).

Some Conclusions

The development of metal recycling

The fact that there is no evidence for the widespread use of recycling until the end of the Middle Bronze Age suggests that this practice was uncommon up to the middle of the 14th century BC. On analysis of the reasons behind this, we can observe how, for example, bronze work in the Argar Culture (1800-1500 BC) only utilised minerals from the region of the production sites (Montero Ruiz 1995: 295). There seems to be evidence in the Early Bronze Age I (2000-1800 BC) in the Central Alps for massive use of the Falhertz mineral, a natural alloy of copper, lead, zinc and silver, which has mechanical properties similar to bronze. Copper-tin alloy was massively introduced later in the Early Bronze Age II (1800-1600 BC) (De Marinis 2006b: 226-227). The fact that documented recycling is started for the first time in the Late Bronze Age (13th century BC) in the Atlantic area of the Iberian Peninsula, and in the Middle Bronze Age (14th-16th centuries BC) in Northern Italy may mean that whilst the early communities were dependent on local resources to make bronze, they did not need to recycle (taking into account that the Iberian Peninsula and northern Italy were rich in tin deposits).

The appearance of copper ingots on a massive scale and biggest in size from large wells may be a common link to the phenomenon of the recycling hoards. The production of raw copper on an almost industrial scale begins when it seems that there is more recycling of bronze.

There is a coincidence chronologically between the "beginning" of the phenomenon of mass scrap collection in the Late Bronze Age and the most prosperous period of trade in the Mediterranean, of greater dynamism in the long distance relationship on continental Europe and some of the major exploitation of copper resources in the Central and Western Alps. This would preclude the use of recycling to make up for a shortage of raw material of copper or tin, for the depletion of mines, or for mobility problems in trade. The most correct interpretation of this phenomenon must therefore be, in our opinion, the emergence of greater accountability in the management of natural resources and the discovery of the trade in scrap bronze.

The problem of correction in alloy and the absence of tin in scrap closets.

Due to the complexity of the chain of production, from mineral extraction to artefacts production, the cycle of metal artefacts is very selective: it can only be undertaken by individuals who possess the secrets in full. The use of recycling could easily lead one to think that this represents a "democratization" of the cycle of metals: it is not necessary to know all the techniques of the cycle, but only those relating to melting. However, the question arises of the difficulty of making tools and weapons from melting recycled scrap, which has acceptable technical properties. One has to be familiar with the art of metalworking to be able to fine tune the metal pool, present in different amounts in different scrap ingots.

In addition to this, one would have to be aware during the collection of scrap of, which fragmented objects have chemical characteristics in the metal alloy that can be advantageous to make a sword, an axe, or a sickle. Linked to the calibration of the copper-tin alloy, there is the almost constant presence of raw copper in hoards with scrap bronze. This may indicate the use of "new" copper to be added to the load of the crucible made from scrap bronze, but also indicates the complete absence of tin in these contexts. Is it preserved in other places? Was it used in such small amounts and so rapidly as to leave no trace in the hoards and in the metallurgical workshop, or scrap hoards? To arrive at some conclusions, one must imagine how the dynamics of a hoard worked: collecting various scrap ingots of bronze over a period of some years, whilst pure copper and tin, were probably sold as scrap ingots, but perhaps to "price" lowest of these. Scrap was then perhaps the most frequently used in the ingots, suffering a major "rotation" in the hoard (which is why there are often scrap objects of various dates). There is the probable choice of the scrap to be recovered and put in storage by the quality of their content; in this case one must consider that the ancient metalworkers possessed the ability to recognize metal content (by colour, weight, or other). Perhaps ingots had a more stable presence in the hoard, because of the use of small amounts to correct a recast of their alloy recycling. If the hoards today are always composed fragments of raw copper alone without any tin, this suggests means that this was used a lot more than copper in the correction and perhaps "rotated" much more rapidly than the copper in hoards.

The recycling of metals can be understood in two types:

1) *Total recycling*: the recasting of one or more scrap objects to create a third new object.
2) *Partial recycling*: "hot", or mechanical working of a damaged object, in order to repair it.

The first type involves the complete destruction of scrap objects, which as a result affects the data that we have today on the production of certain types of objects, as well as, in the case of a long-range movement of scrap, on the productivity of a region. It is also indicative that most categories of objects that can break are found in the scrap hoards, that is tools and weapons, but almost never ornaments. Artefacts that are recycled are, in fact, subject to a break. This means, that they always exert influence on the data that we have. Thus the production of weapons and tools is greatly overestimated compared to that of ornaments. Related to this type of recycling, there is the issue of the economic and social role of those who organised the movement of scrap. Given the large amount of evidence for recycling in the Late Bronze Age, it is certain that the scrap market sustained much of the bronze production in Europe from the 13th to the 8th / 7th century BC. It is likely that most of Europe was organised in a system of "chiefdom" societies, with a large part of the social prestige and economic power of local leaders based on the management of scrap bronze, as well as on the trade in exotic materials (amber, glass) and the management of war, which is probably also linked to the "production" and need to recast scrap bronzes.

The second type was adopted to a large extent with the introduction of iron, because the technology of casting iron was not known until the Middle Ages. However, one may envisage its use for copper alloys in Bronze Age. In fact a broken sword blade could possibly be remodelled to make a dagger, as demonstrated by some examples of the Iberian Peninsula in the Late Bronze Age (Goiachina *et al.* 2008: 164).

Bibliography

ANTONACCI SAMPAOLO, E., CANZIANI RICCI, C. and FOLLO, C.
1992 Il deposito di S. Francesco (Bologna), In *Archeometallurgia. Ricerche e prospettive*. Atti del Colloquio Internazionale di Archeometria, Bologna: Edizioni CLUFB, pp. 159-206.

BARGE, H.
2004 Le dépôt de bronzes de Moriez (Alpes-de-Haute-Provence). *Documents d'Archéologie Méridionale* 27: 141-170.

BARTOCCI, A. and MARIANESCHI, E.
1960 *I metalli e l'acciaio*. Terni: Polifrafico Alterocca

BIETTI SESTIERI, A. M.; GIARDINO, C.
2003 Alcuni dati sull'industria metallurgica in Abruzzo, In *Atti della XXXVI Riunione Scientifica dell' Istituto Italiano di Preistoria e Protostoria nelle Abruzzo*, Firenze: Istituto Italiano di Preistoria e Protostoria, pp. 411-430.

BOCQUET, A. and LEBASCLES, M. C.
1983 *Metallurgia e relazioni culturali nell'età del Bronzo Finale delle Alpi del Nord Francesi*, Torino: Antropologia Alpina.

BRANDHERM, D.
2007 Sobre el origen del fenómeno de los depósitos en la Península Ibérica: ocultaciones de objetos metálicos de los inicios de la Edad del Bronce, In Celis Sanches, J., Delibes De Castro, G., Fernández Manzano, J. and Grau Lobo, L. (eds) *El hallazgo leonés de Valdevimbre y los depósitos del Bronce Final Atlántico en la Península Ibérica*, León: Junta de Castilla y León/ Diputación de León, pp. 176-193.

BRADLEY, R.
1988 Hoarding, recycling and the consumption of prehistoric metalwork: technological change in Western Europe, *World Archaeology* 20: 249-260.

CARANCINI, A. and PERONI, R.
1997 La *koiné* metallurgica, In Bernabó Brea, M., Cardarelli, A. and Cremaschi, M. (eds) *Le Terramare. La piú civiltà padana*, catalogo della mostra di Modena, Milano: Electa, pp. 595- 601.

CARDOSO, J. L. and VILAÇA, R.
2008 Artefactos da Idade do Bronze na região de Chaves, *Revista Portuguesa de Arqueologia*, 11(3): 41-54.

CIERNY, J.
1997 Rame, stagno e bronzo, In Endrazzi, L. and Marzatico, F. (eds.) *Ori delle Alpi*, Catalogo della mostra di Trento, Castello del Buonconsiglio, Trento: Provincia Autonoma di Trento, 75-82

CIERNY, J., MARZATICO, F., PERINI, F., and WIESGERBER, G.
1998 Prehistoric copper metallurgy in the southern alpine region, In Mordant, C., Pernot, M. and Rychner, V. (eds) *L'atelier du bronzier en Europe du XXe au VIII^{e}avant notre ère. Du mineral au métal, du métal á l'object*, Actes du colloque international "Bronze '96", de Neuchatel et Dijon 1996, tome II, Paris: Comité des travaux historiques et scientifiques, pp. 25-34.

COFFYN, A.
1985 *Le Bronze Final Atlantique dans la Péninsule Ibérique*. Paris: Diffusion de Boccard.

CONATI BARBARO, C.
2005 Scelte tecnologiche e identitá culturali: alcune riflessioni, *Origines* XXVII: 171-190. Roma: Universitá degli Studi La Sapienza.

COURTOIS, J. C.
1960 Les dépôts de fondeur de Vernaison (Rhône) et de la Poype-Vaugris (Isère), *Cahiers Rhodaniens* 7: 3-24.

DE MARINIS, G.
2005 Tecnologie produttive nei siti dell' età del Bronzo di Moscosi di Cingoli e Cisterna di Tolentino, In *Atti della XXXVIII Riunione Scientifica dell' Istituto Italiano di Preistoria e Protostoria nelle Marche*, Firenze: Istituto Italiano di Preistoria e Protostoria, pp. 679-694.

DE MÁRINIS, R. C.
1997 Il ripostiglio della Baragalla, In BERNABÓ BREA, M.; CARDARELLI, A.; CREMASCHI, M. (eds) *Le Terramare. La piú civiltá padana*, catalogo della mostra di Modena, Milano: Electa, 306-308

2006a Circolazione del metallo e dei manufatti nell'età del Bronzo dell' Italia Settentrionale, *Atti della XXXIX Riunione Scientifica dell' Istituto Italiano di Preistoria e Protostoria*, Firenze 25-27 novembre 2004, Firenze: I.I.P.P., 1289-1317.

2006b Aspetti della metallurgia dell'età del rame dell' antica Età del Bronzo nella Penisola Italiana, *Rivista di Scienze Preistoriche*, LVI, Firenze: Istituto Italiano di Preistoria e Protostoria, 211-272.

DEL LUCCHESE, A. and DELFINO, D.
2008 Metallurgia protostorica in Val Bormida, *Archeologia in Liguria*, n.s., 1, 35-47

DELIBES DE CASTRO, G.
1997 Una introducción al tema de los depósitos del Bronce Final en el Oeste de Europa, *Acontia*, 3, Valladolid, 61-72

DOMERGUE, C.
2004 Les mines et la production des métaux dans le monde méditerranéenne au Ier Millénaire avant notre ère. Du producteur au consommateur, In LEHOERF, A. (ed.) *L'artisanat métallurgique dans les sociétés anciennes en Méditerranée Occidental*, Paris: Collections de l'Ecole Française de Rome, vol. 332, 129-159

DORO, A.
1975 Un ripostiglio di bronzi a Pinerolo, *Sibrium*, 12, 205-222

FASNACHT, W.
1997 Tecnologia del bronzo, In ENDRAZZI, L., MARZATICO, F. (eds) *Ori delle Alpi*, Catalogo della mostra di Trento, Castello del Buonconsiglio, Trento: Provincia Autonoma di Trento, 83-88

GAMBARI, F. M., and VENTURINO GAMBARI, M.
1994 Produzioni metallurgiche piemontesi: la Tarda età del Bronzo, *Quaderni della Soprintendenza Archeologica del Piemonte*, 12: 23-41.

GARCIA, D.
2003 Les dépôts d' objects en bronze protohistoriques en Provence-Alpe-Côte d' Azur: un état de la question,*Documents d'Arquéologie Méridionale*, 26, 377-384

GIARDINO, C.
2005 Metallurgy in Italy between the Late Bronze Age and Early Iron Age: the coming of iron. In Attema, P., Nijboer, A. and Zifferero, A. (eds) *Papers in Italian Archaeology VI. Communities and Settlements for the Neolithic to the Early Medieval Period*, Proceeding of 6thConference in Italian Archaeology held at the University of Groeningen, Groeningen Institute of Archaeology, The Nethelands, April 15-17, 2003, Volume I, Oxford: Archaeopress, BAR International Series 1452 (I): 491-505.
2010 *I metalli nel mondo antico. Introduzione all' archeometallurgia*, Roma-Bari: Laterza Editore.

GOIACHINA, J., GOMEZ DE SOTO, J., BOURCHIS, J. R.; and VEBER, C.
2008 Un depôt de la fin de l'Age du Bronze à Mechers (Charente- Maritime), *Bulletin de la Societé Prehistorique Française*, tome 105 (1): 159-185.

HARDING, A. F.
2000 *European societies in the Bronze Age*, Cambridge: Cambridge University Press.

JALLAY, E.
1944 O esconderijo pré-histórico de Porto do Concelho (Mação, Beira Baixa), *Brotéria*, Porto, XXXVIII, 263-277.

KALB, P.
1978 Senhora da Guia, Baiões. Die ausgrabung 1977 auf einer Hohensiedlungder Atlantishen Bronzezeit in Portugal, *Madrider Mitteilungn* 19: 112-138.

LEMONNIER, P. (ed)
1993 *Technological choices: transformation in material culture since the Neolithic*. London: Routledge.

LO SCHIAVO, F.
2006 Ipotesi sulla circolazione dei metalli nel Mediterraneo centrale, *Atti della XXXIX Riunione Scientifica dell' Istituto Italiano di Preistoria e Protostoria*, Firenze 25-27 November 2004, Firenze: I.I.P.P., 1319-1337.

MANNONI, T and GIANNICHEDDA, E.
1996 *Archeologia della produzione*. Torino: Einaudi.

MARINO, D. and PIZZITUTTI, G.
2008 Un ripostiglio di bronzi del territorio a sud di Crotone (Calabria centro orientale), *Rivista di Scienze Preistoriche*, LVIII: 321-336, Firenze: Istituto Italiano di Preistoria e Protostoria.

MARZATICO, F.
1997 L' industria metallurgica nel Trentino durante l'etá del Bronzo, In Bernabó Brea, M., Cardarelli, A. and Cremaschi, M. (eds) *Le Terramare. La piú civiltá padana*, catalogo della mostra di Modena, Milano: Electa, 570-591.

MONTERO RUIZ, I.
1995 *El origen de la metalurgia en el sudeste de la Penisula Iberica*, Almeria: Instituto de Estudios Almerienses.

MONTERO RUIZ, I., HUNT ORTIZ, M. and SANTOS ZALDUEGUI, J. F.
2007 El depósito de la Ria de Huelva: procedencia del metal a través de los resultados de análisis de isotopos de plomo, In Celis Sanches, J., Delibes De Castro, G., Fernández Manzano, J. and Grau Lobo, L. (eds.) *El hallazgo leonés de Valdevimbre y los depósitos del Bronce Final Atlántico en la Península Ibérica*, León: Junta de Castilla y León/Diputación de León, pp.196-209.

MORDANT, C.
2003 Les depôts d'objects métalliques de l'âge du Bronze dans l'Est de la France. Nouvelles approches et méthodes d'études, *Documents d'Arquéologie Meridionale*, 26: 371-376.

MUTI, A.
1993 *Caratteristiche e problem del popolamento terramaricolo in Emilia Occidentale*, Bologna: University Press.

NEEDHAM, S. P.
1998 Modelling the flow of metal in the Bronze Age. In Mordant, C., Pernot, M. and Rychner, V. (eds) *L'atelier du bronzier en Europe du XXe au VIII^{e}avant notre ère. Du mineral au métal, du métal á l'object*, Actes du colloque international "Bronze '96", de Neuchatel et Dijon 1996, tome II, Paris: Comité des travaux historiques et scientifiques, 285-301.

PIPINO, G.
2008 Risorse metallifere storiche nelle alti valli della Bormida: giacimenti cupriferi di Murialdo, Bormida e Mallare, In Del Lucchese A., Gambaro, L. (Eds.) *Archeologia in Liguria*, n.s., I, 2004-2005, Genova: Editore De Ferrari, 48-58.

ROSSI, M. and GATTIGLIA, A.
1998 I recenti scavi a La Croupe de Casse Rousse (Hautes Alpes) e il concetto di ripostiglio. In De Marinis, R. C., Bietti Sestieri, A. M., Peroni, R. and Peretto, C. (eds) *Proceeding of XIII International Congress of Prehistoric and Protohistoric Sciences*, Forlí: Abaco, 321-327.

SALZANI, L.
2000 Fratta Polesine: il ripostiglio di bronzi n.º 2 da Frattesina, *Quaderni di Archeologia del Veneto*, XVI: 38-46.
2003 Fratta Polesine: il ripostiglio 4 e altri reperti da Frattesina, *Quaderni di Archeologia del Veneto*, XIX: 40-45.

SENNA-MARTINEZ, J. C.
2007 Depósitos "Versus" Oficinas de Fundidor: Problemas contextuales de la Arqueometalurgia en Portugal. In: Grau Lobo, L. (ed.) El Hallazgo Leonés de Valdevimbre y los Depósitos del Bronce Final Atlántico en la Península Ibérica. León: Museos de Castilla y León. «Estudios y Catálogos» 17, pp. 258-279.

VILAÇA, R.
1997 Metalurgia do Bronze Final da Beira Interior, *Estudos Pré-Históricos*, 5: 123-154.
2006 Depósitos de Bronze do Território Português. Um debate aberto, *O Arqueólogo Português*, IV(24): 9-150.

Rock Art Recycled?
On the Use of Bronze Age Rock Art Sites during the Iron Age in Southern Scandinavia[1]

Per Nilsson

Abstract

In recent years a number of excavations at rock art sites have been conducted in southern Scandinavia. The rock art in this region is mainly dated to the Bronze Age, but many of the finds and features found at these excavations have been dated to the Iron Age. This raises some questions: How did people during the Iron Age relate to rock art sites? Were they regarded as pictures from an ancestral past, were they abandoned or perhaps simply forgotten? It is proposed here that the finds and features found in close connection to rock art sites can be seen as the material remains of a dialogue with the past. It is also suggested that in a time of societal change the past can become the 'Other'. I will give some examples of reused rock art sites from different parts of southern Scandinavia, with a focus on the rock art region of Himmelstalund in the southeastern part of Sweden.

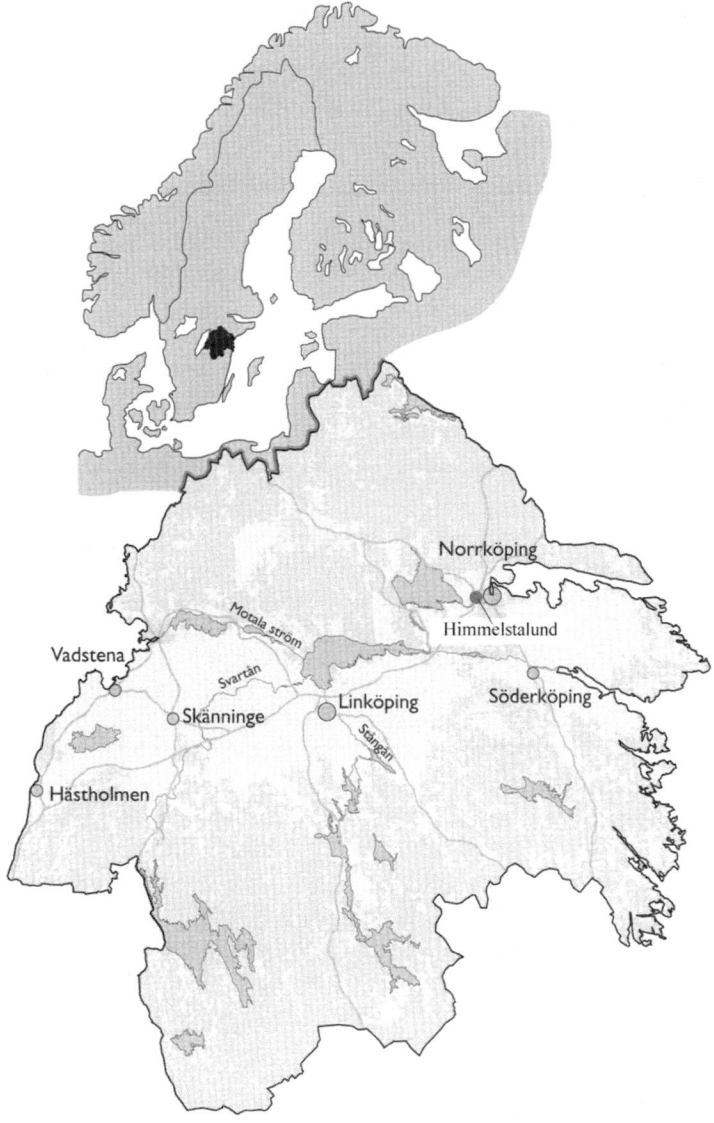

Fig. 1. *Map showing the location of the Himmelstalund region.*

---

[1] This paper was presented and written in connection with the EAA-conference in Riva del Garda in 2009. Since then a number of excavations at rock art sites in Scandinavia have been conducted and published. The results from these later excavations are not included in this text.

Introduction

In the spring of 2007 a small excavation was conducted beneath one of the major panels at the rock art site of Himmelstalund, situated west of the town of Norrköping in the south-eastern part of Sweden (Figs 1, 4). Two hearths were found in close connection to one of the panels and they were both C14-dated to the Early Iron Age. The figurative rock art in this region has been dated to the Bronze Age, so the dating of the hearths gave rise to some questions: If the hearths had been dated to the Bronze Age the connection between them and the panel had seemed pretty clear. What, then, was the role of the rock art sites during the Iron Age? Were they abandoned, reused or perhaps simply forgotten?

One of Scandinavia's densest concentrations of rock art is to be found around the now regulated rapids in the river Motala Ström, west of the town of Norrköping. The best known site is Himmelstalund, with some 60 panels featuring more than 1,700 figures (Selinge 1985: 110f; Nilsson 2008a). What makes this region special is the spatial connection between the rock art sites and contemporary cemeteries, such as Fiskeby, Ringeby and Klinga and the nearby settlement at Pryssgården (Lundström 1965, 1970; Stålbom 1994; Kaliff et al 1995; Borna-Ahlkvist et al 1998). During the last decade several dissertations on different aspects of the Bronze and Early Iron Age societies in this region have been published, as well as more detailed studies and excavation reports (Broström and Ihrestam 1993; Kaliff 1997; Stålbom 1998; Borna-Ahlkvist 2002; Wahlgren 2002; Coles 2003; Fredell 2003; Helander 2005; Ericsson and Nilsson 2007; Nilsson 2005a, 2005b, 2007, 2008a, 2008b; 2010a, 2010b; Tilley 2008). Although the rock art at the Himmelstalund site was discovered during the first half of the 19th century, the first scientific investigation did not take place until 1871 (Nordén 1925). In 1903 it was documented by Oscar Almgren and the then Crown-Prince Gustav Adolf, and during the 1910-20's more thoroughly by Arthur Nordén. The rock art at Himmelstalund has been documented on several occasions since then (Burenhult 1973, 1980; Selinge 1985). But Nordén's dissertation, "Östergötlands bronsålder" [Östergötland´s Bronze Age], published in 1925, is still the most comprehensive study of the county's – and especially the Norrköping area's – rock art.

Excavations at Rock Art Sites

In recent years there has been an increasing interest in performing excavations at rock art sites in Scandinavia (Bengtsson 2004; Goldhahn 2006; Kaul 2006). The south Scandinavian rock art tradition has mainly been dated to the Bronze Age, and several of the finds and features discovered at these excavations can be dated to this period. But there are also a significant number of finds and features from later periods, mainly from the Early Iron Age. In the following I will present some examples of excavations at rock art sites in southern Scandinavia, beginning with excavations from the Himmelstalund region.

Only a few excavations at rock art sites have been conducted in the Himmelstalund region since Arthur Nordén's ground breaking research in the 1910-20's. When Nordén was searching for new rock art he had to remove the existing topsoil in many cases, in order to find the figures. In quite a few cases he noted that the motifs were covered with a layer of fire cracked stone, charcoal and soot. He also excavated several heaps of fire cracked stones (burnt mounds), at some of the rock art sites. The main aim for Nordén though, as well as for most of the rock art scholars working in this area, has been to undertake new documentation of earlier known sites and/or discover new rock art. Later excavations with the expressed ambition to study the relationship between rock art sites and contemporary (or later) finds and features have been rare (cf. Lødøen 2006: 5).

The Himmelstalund site

Nordén noted that some of the panels at the Himmelstalund site were covered by a layer of fire cracked stones and black sandy soil and that several of the figures had been damaged by fire (Nordén 1925: 48f). When the panel beside the two Iron Age hearths (Fig. 4) was documented in the beginning of the 20th century, at least a part of it was covered by a thick layer of soil (Prince Gustav Adolf 1904). Unfortunately it is not possible to determine whether this layer also contained fire-cracked stones. About 100 meters to the north of the rock art site some hearths, post holes and cultural layers have been discovered (Persson 1998). Two of the features, a hearth and a small posthole, were dated to the transition between the Late Neolithic and the Early Bronze Age (1920-1740 BC, 1740-1530 BC, Cal 1 Sigma).

In recent years the Swedish National Heritage Board has carried out archaeological excavations at Himmelstalund (Ericsson and Nilsson 2007; Broström 2007; Nilsson 2008a-b). A large three-aisled house (circa 7.5 x 3 m) was found on a natural ledge between the rock and the river Motala Ström (Figs. 2-3). The house consisted of three pairs of trestles and two pits filled with burnt clay were found along its walls. A row of post holes, which might have signified a confined area between the rock and the river, were found at the western gable end of the house. Charcoal from one of the post holes inside the house was dated to 45BC-25AD (Cal 1 sigma). Settlement remains were also discovered some 100 metres to the west of the rock art site. Five 14C-dates indicated that this settlement had been in use between 350BC and 235AD (Cal 1 sigma).

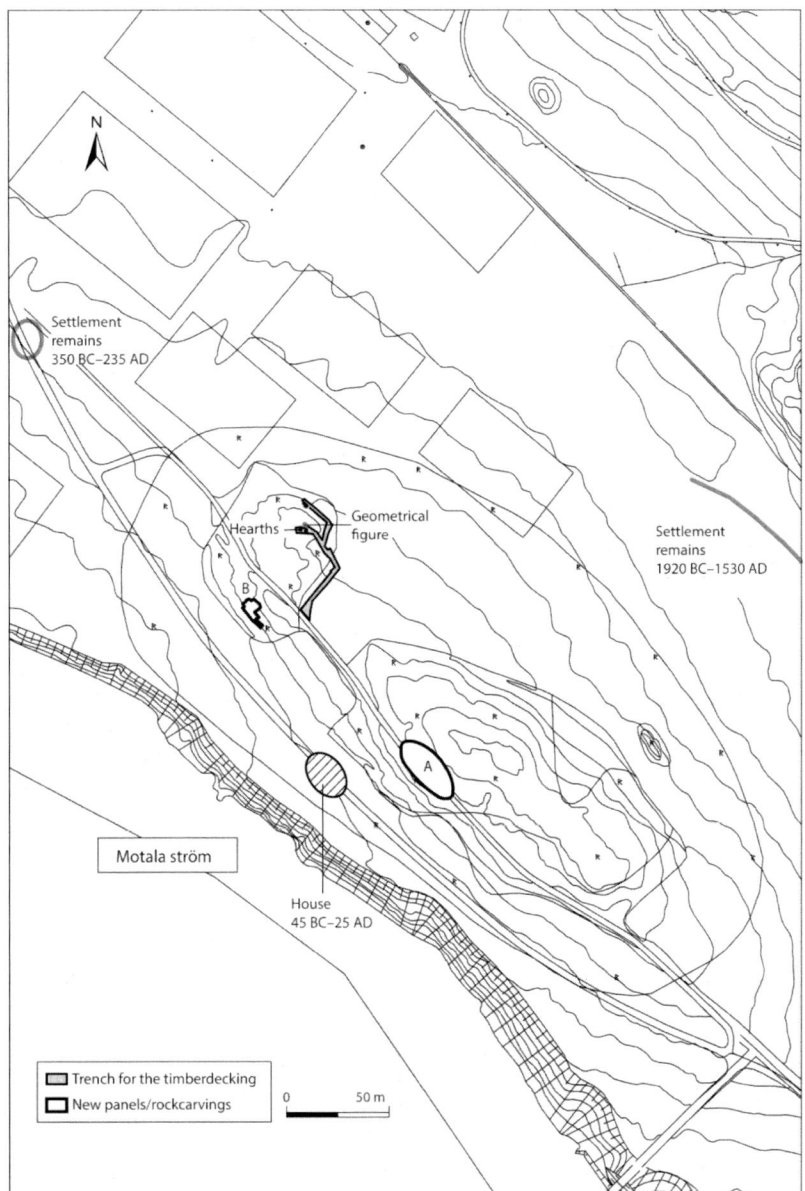

Fig. 2: *Map showing the rock art site at Himmelstalund with the location of the hearths and the settlement remains.*

The two hearths mentioned in the beginning of the article were found about a metre or so away from one of Himmelstalund's best known motifs (Fig. 4); a deeply carved geometrical figure (or perhaps a net?) associated with a human carrying a spear. The two hearths were dated to the Pre-Roman and Early Roman Iron Age (90BC-AD20, AD135-230, Cal 1 sigma). One question that arises relates to the connection of the two hearths with the adjacent rock art panel. Were they connected in any way, or had the motifs been forgotten or lacked significance when the fires by the rock were lit? The time interval between the two hearths is too great for them to be contemporary with each other, so the hearths belong to two chronologically separate occasions. As the two hearths are both contemporary with the settlement remains to the south and west of the rock it is reasonable to suggest that they can be associated with activities carried out in connection with these settlements, regardless of whether or not the activities were of a ritual or more everyday nature.

The oldest rock art figures in the Norrköping area consist of weapons such as swords and axes, and the oldest swords should probably be dated to the transition between Early Bronze Age Period I and II, i.e. ca. 1600 BC (Wahlgren 2002: 178, 254). If this is accurate these settlement remains could be contemporary with, or perhaps even older than the oldest rock art motifs at Himmelstalund. Nevertheless, it is interesting to note the existence of a few remains from a possible Late Neolithic/Early Bronze Age-settlement, as well as a more substantial settlement from the Early Iron Age. But what we have not yet found at Himmelstalund are finds or features from later parts of the Bronze

Fig. 3: *A house from the Early Iron Age was found between the rock art site and the nearby river Motala Ström. Photo Per Nilsson.*

Fig. 4: *Two hearths were found beneath one of the panels at Himmelstalund. Photo Per Nilsson.*

Age, when the rock art activity was more intense (e.g Wahlgren 2002: 238f). Perhaps a settlement from this period is yet to be found, but it is also possible that the Himmelstalund site is yet another example of a major rock art site with no or few finds and features that are contemporary with the rock art (Goldhahn 2006: 92).

Leonardsberg

Traces of fire related activities such as layers of fire cracked stones covering rock art and figures damaged by fire were found by Arthur Nordén at Leonardsberg - another major rock art site situated about a kilometre to the west of the Himmelstalund site (Fig. 5). Only one excavation has been performed at this site since Nordén's excavations in the 1920's. A few years ago a burnt mound beside one of the panels was excavated (No 29, Wahlgren 2002). During the excavation the burnt mound turned out to be a grave from the Late Iron Age. Five hearths were found under the grave and were interpreted as part of a larger system of hearths. Two of the hearths were dated to the Roman Iron Age (AD240-340 and AD206-410, Cal 1 sigma). But a layer of soot, situated on top of the panel, was dated to the Late Bronze Age (900-800BC, Cal 1 sigma). It is interesting to note that the site has been repeatedly used since the Bronze Age up to the Late Iron Age.

Fiskeby

At Fiskeby a large cemetery with more than 500 burials was excavated in the 1950's (Lundström 1965, 1970). The cemetery was established in the Late Bronze Age and continued in use until the Late Iron Age. What is especially interesting about this cemetery is that it was located just beside a rock art site. The excavators were explicitly searching for rock art during the excavation, but only one of all the graves covered a rock art panel. They also noted that no graves were placed on the rocky surfaces that divided the cemetery. When more graves were added during the Iron Age, the cemetery expanded away from the panels (Lundström 1970: 116). It therefore seems likely that the motifs were known and in a sense also respected during the Bronze Age as well as during the Iron Age (Fig. 6).

Rock Art Excavations from Other Regions

There are in fact other examples of possible reuse of rock art sites from different parts of southern Scandinavia. In a study from 2005 Lasse Bengtsson presents a list of 30 excavations at rock art sites in Sweden and Norway (Bengtsson 2004; c.f Nordström 1995). The list below is based on both Nordström's and Bengtsson's studies and it only includes finds and 14C-dates from periods

Fig. 5: *Rock art covered with fire cracked stones at Leonardsberg. From Nordén 1925.*

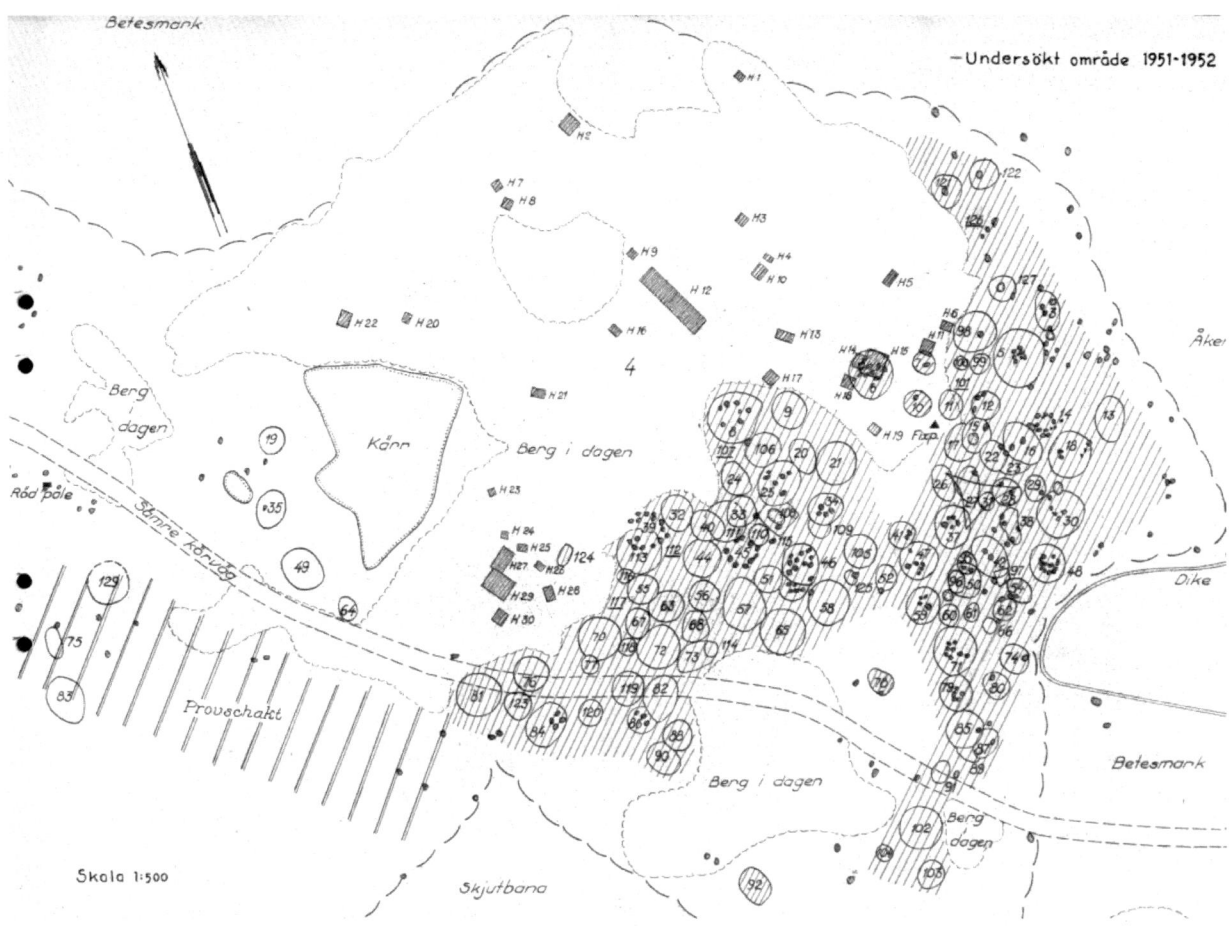

Fig. 6: *The cemetery at Fiskeby. Map by Per Lundström, copy from the Royal Swedish Academy of Letters, History and Antiquities.*

later than the Bronze Age. Bengtsson's survey is based on different categories of finds or features, and the examples below are therefore presented in the same way:

Potsherds were found at 18 of the 30 excavations. The sherds that were found at the excavations in the county of Bohuslän were all dated typologically to the Early Iron Age. However it is important to note that it can be hard to separate Late Bronze Age ceramics from Early Iron Age ceramics. At three of the sites; Drottninghall (Scania), Kalleby (Tanum in Bohuslän) and Hällby (Uppland), sherds were also found that should probably be dated to the Early Iron Age/Migration Period (Nordström 1995: 31).

Stone pavings/constructions were found at 14 of the locales. Two carbon samples from a stone paving in Högsbyn in Dalsland (No. 11) were dated to the Pre-Roman Iron Age/Early Roman Iron Age (350 BC–AD60, Cal 1 sigma) and the Early Medieval period (AD1060-1400, Cal 1 Sigma). It proved possible to date two areas of stone paving in Bohuslän. Both were dated to the Iron Age; No. 1371 in Tanum (390-170 BC) and No. 446 in Tossene (50 BC – AD180).

Hearths and layers of fire cracked stones were found at nine of the 30 locales. At one of the sites, No. 897 in Tanum, a hearth was dated to 200BC–AD130. A bead from the Roman Iron Age was found in a layer of fire-cracked stones at the same site. Bengtsson suggests that the layer could have been placed by the rock art more by chance, but also points out the possibility of a contextual relationship between the rock art and the layer of fire cracked stone during the Roman Iron Age. What is interesting is also that only one of the hearths found at rock art sites in Bohuslän has been dated to the Bronze Age, according to Bengtsson (2004). Another example is from Drottninghall in Scania where a hearth was found beneath one of the panels. The hearth was C14-dated to AD560-780 (Cal 1 Sigma). Layers of fire cracked stone were found at 12 of the locales. Bengtsson notes that there seems to be a clear connection between rock art and fire related activities such as hearths and fire cracked stone.

Further excavations have been conducted during recent years at the Tossene site in Bohuslän (Bengtsson and Ling 2007). There are many interesting results from these excavations, but what is especially important for my own concern is the possible reuse of the site during

Fig. 7: *At Lille Strandbygård on the Danish island of Bornholm, two houses were found beside the rock art. From Sørensen 2006: 72.*

the Iron Age. At one of the panels there is a fascinating motif of a warrior caught in an acrobatic pose and in front of this panel two hearths were found. The hearths were dated to the Late Roman Iron Age (AD250-410 and AD250-420, Bengtsson and Ling 2007: 45). The similarity between the latter site and the hearths found at Himmelstalund is striking. In recent years a number of excavations at rock art sites have been carried out on the Danish island of Bornholm (Kaul 2005, 2006; Sørensen 2006). At Madsebakke a cultural layer at the 'entrance' to the panel was dated to the Early Roman Iron Age (AD0-150). (Fig. 7)

Potsherds from the same period were also found. During the later part of the Iron Age, about 200-600 AD, a cemetery was established on top of the rock at Madsebakke. A house structure from the Late Iron Age was also found in close connection to the rock art. House structures were also discovered at a minor excavation at Lille Strandbygård, on the southern part of the island. Here, two successive house structures dated to the Iron Age were found just beside the rock art. According to the excavators, one of the possible entrances to the house(s) was most probably placed just by the decorated panel.

Relating to Rock Art Sites during the Iron Age

The examples above have shown that a spatial relationship exists in many cases between rock art sites and later remains, such as barrows, cemeteries and settlements. However, is it possible to determine whether this spatial relation also reflects a contextual relationship? In some of the above mentioned examples the excavators were convinced of a contextual relationship between the rock art and the adjacent remains during the excavation. However, when the C14-dates arrived the idea of such a contextual relationship was questioned. To determine whether the relationship is contextual and not only spatial is of course always a challenge and it is important that each site is studied within its own specific context. But I do believe that we should be more open to the possibility of the existence of such a relationship, especially when it comes to excavations at rock art sites. In the following I will present some more examples of possible reuses of rock art sites during the Iron Age:

Rock art and Iron Age graves

In a dissertation from 1987 Ulf Bertilsson states that the location of Iron Age graves on or adjacent to rock art panels in the county of Bohuslän was a conscious act (Bertilsson 1987: 149). This relationship is also discussed by Bengtsson and Ling (2007: 48), who admit that they are a bit puzzled by the strong reminiscence of Iron Age activity at rock art sites. The disappearance of the figurative rock art coincides with the first general impact of agriculture in Bohuslän. Bengtsson and Ling (2007: 48-49) regard the rock art sites as reflections of mobile/seasonal activities during the Bronze Age, and propose that this system was altered during the Iron Age when a more agrarian system was established. They suggest that the rock art sites were then reused or revitalized within the agrarian tenure system, although no new (figurative) motifs were made. There are several examples from the Himmelstalund region of what can be interpreted as a conscious location of grave mounds close to rock art sites, for instance at Borgs Säteri (No. 16-17) where there is a complex relation between Iron Age barrows, burnt mounds and rock art panels (Nordén 1925: 82f; Wahlgren 2002: 149). It is worth noting that several of the barrows classified as burnt mounds turned out to be Iron Age graves after excavation. To add some complexity to this, there are also examples of burnt mounds from the Bronze Age/Early Iron Age that have been reused as graves during the Late Iron Age (Nordén 1925: 83).

Iron Age rock art

The rock art tradition in the Himmelstalund region has been dated to the Bronze Age, with a possible continuation throughout the Pre-Roman Iron Age (Wahlgren 2002: 179). Perhaps a few of the motifs in the Himmelstalund region can be dated to the very Late Bronze Age or Early Pre-Roman Age. But a brief survey of the panels documented by Nordén shows few chronological traits from this period, such as horse riders with rectangular shields or ships with bifurcated stems (Kaul 2004: 310, 394f). In other areas, such as Tanum in Bohuslän, Bornholm and Trøndelag, there are several examples of ship figures that can be dated to the Pre-Roman Iron Age (Kaul 2004: 394f). Johan Ling has recently shown that at least 130 of the ship figures from the parish of Tanum could be classified as ships from the Pre-Roman Iron Age (Ling 2008: 196). If this dating is correct, it is also conceivable that other, older motifs could have been recarved and revived during this period.

Runic inscriptions on rock art panels

An interesting example of an early approach (or dialogue?) with the rock art at Himmelstalund is the line of runic inscriptions carved on the same rock as the geometrical figure, some metres to the northwest of the two previously mentioned hearths (Nordén 1925, 1936). The line of runes consists of 5-6 (reversed) characters of the older *futhark* and has been interpreted in several different ways (Fig. 8).

The most likely interpretation is *Braido* or *Brajdo*, which could refer to a woman's name meaning "the wide or the broad" (Svante Lagman and Patrik Larsson, runologists, personal communication 2007). Another alternative is *Buajdo*, i.e. "I did". The authenticity of the runes has been discussed, but if genuine they could have been carved as early as 200-300 AD – although a later dating to 400-500 AD is also reasonable. Other examples of runic inscriptions on rock art panels can be found in Tanum and Utby on the Swedish West Coast, as well as in Kårstad in Norway (Gerdin and Munkenberg 2005; Mandt 2005; Ling 2008: 72-73).

Fig. 8: *The runic inscription from Himmelstalund. Photo Per Nilsson.*

Rock art and traces of fire

Traces of fire related activities have been found at many rock art sites. In the parish of Askum in Bohuslän no less than 71 of the 247 sites showed traces of fire damage (Bengtsson 2004: 37). In some cases rock art has even been destroyed by fire. Whether this damage should be regarded as contemporary with the rock art or not can be hard to determine. One possible explanation for some of these traces of damage is that they are derived from Bronze Age cremation ceremonies (Goldhahn 2007: 261f). But the number of fire related activities dated to the Iron Age at rock art sites indicates that some of the damage could have been caused during this period too. One specific topic is the relation between rock art and layers/heaps of fire-cracked stones (burnt mounds). Nordén found traces of fire related activities at many of the rock art sites he investigated; for example at Leonardsberg, Himmelstalund, Skälv, Egna Hem and Borg (Nordén 1925, c.f. Wahlgren 2002: 258). At one of the sites he excavated a large burnt mound that partly covered a panel with rock art. During the excavation he found a button from the Late Bronze Age. For Nordén this was a clear indication that many of the burnt mounds could be dated to the Bronze Age (No. 46, Nordén 1925: 64). During the Early Iron Age there seems to be an emphasis on different aspects of fire, with hearths and other features being found in large numbers, not just at rock art sites but also at settlements and cemeteries. One alternative is that the hearths found at rock art sites ought to be regarded as remains from shepherds' camps. This has recently been suggested as a possible explanation for the large number of isolated hearths dating back to the Late Bronze- and Early Iron Age (Petersson 2006: 169). I think it is reasonable to interpret the main part of the hearths and layers/heaps of fire cracked stones at rock art sites as consciously located (cf. Kaliff 2007: 105). There is of course also the possibility that some of the hearths were lit by shepherds. The large scale herding system was established during the Bronze Age and it is therefore likely that the rock art sites located within the grazing areas were connected with special beliefs or perhaps myths during the Iron Age too.

Towards a More Problematic Past

The south Scandinavian Bronze Age societies have most often been interpreted as hierarchic and stratified (Kristiansen and Larsson 2005). In this society the older (monumental) structures, such as barrows, cairns and major rock art sites played an important role for the reproduction of the existing society. But what role did these monumental places play when the Bronze Age culture came to an end? During the Early Iron Age the thousand year old tradition of making figurative rock art ended. How long did the memory of this tradition and the places connected with it remain in the minds of the people during the Iron Age?

Since the early 1990's there has been an increasing interest in discussing how and why monumental remains were used and/or reused during later periods, a field of research also known as "The past in the past" (Bradley 2002). This field can also be seen as a subdivision of a broader discussion concerning different conceptions of time within archaeological and anthropological research (Lucas 2005). Another subdivision is the biographies, or life histories of certain types of prehistoric remains, mainly different kinds of monuments (e.g. Holtorf 2000-2007, 2008). A number of articles and monographs have been published and the history of research on the subject is now quite extensive (see Bradley 2002; Lucas 2005; Jones 2007; Thäte 2007). In Sweden, the discussion has mainly focused on the way barrows or burial grounds were used or reused during later periods (e.g. Artelius 2004; Munkenberg 2004; Larsson 2005; Strömberg 2005; Artelius and Lindqvist 2007; Tegnér 2008; Thäte 2007). The reuse of, that is deposition/sacrifice at older graves and settlements have most often been interpreted as a legitimizing act, as a way of demonstrating the rights to use or inhabit a certain area (Thäte 2007: 31f). But these places could also have been regarded in a different way. There are some interesting discussions of other, perhaps more problematic views of the past. One example is Borna-Ahlkvist's discussion of how older house ruins could be regarded as still inhabited by the feared spirits of the ancestors (Borna-Ahlkvist 2002: 189; cf. Gerritsen 1999; Holtorf 1999: 443f; Tegnér 2008; Tilley 2008: 251f). Another view is given by James Whitley in the article "Too many ancestors", where he gives a critical response to the widespread use of the ancestors in explaining why monuments were reused in Britain. He suggests that we should look at alternative hypothesis for the reuses and depositing at monuments, which includes the offerings to and worshipping of aliens and previous races Whitley (2002: 124). Another interesting point of view is presented by Andrew Jones. In an essay called "Memory and Material Culture" he discusses different aspects of how we remember (Jones 2007). One of his points is that material culture could be viewed as mnemonic traces, and as such act as aids for remembering. What is especially interesting from my point of view is Jones discussion of indexicality:

> "Each event, whether the production of an artefact, its deposition, the act of building monuments, or the act of inscription, is an index related to other events as part of an indexical field" (Jones 2007: 226).

These indexes can also relate to past events, and I believe that the deliberate placing of for example Iron Age graves and layers/heaps of fire cracked stones on top of or beside the rock art could be interpreted as such an indexical relation to the past.

The 'Other'

During the Bronze Age as well as during the Early Iron Age, a number of cemeteries and settlement sites are established, while others are abandoned (Petersson 2006: 20ff). This holds true for many of the rock art sites as well. Not all of them were used during the whole of the Bronze Age, or at least no new figures were made. There are also examples of rock art figures where additions have been made during the course of the Bronze Age (Fredell 2003: 229). How are we to understand these abandoned places and remade figures? Were some of them connected, or loaded, with negative connotations? Could the deliberate covering of rock art panels with layers and heaps of fire cracked stones be interpreted as an example of a more problematic relation to the past? One way of discussing this relation to the past is to consider if the past sometimes could have been regarded as something unwanted, or at least as something one no longer wanted to be associated with. This can also be described with a philosophical term - the 'Other'. The use of the 'Other' as a philosophical concept has a long history and it was used already by Hegel, in describing how ones "Self" is always constituted by the 'Other'. Within postcolonial theory the concept has been used to describe how the West treated, or understood, the people in the colonized countries (e.g. Fanon 1962; Said 1978; Bhabha 1994). With influences from postcolonial theories, the concept has also been used within Swedish archaeological research. One example is a discussion regarding cultural meetings and different uses of the concept of the 'Other' by Cornell and Fahlander (2002: 27). The term has also been used to discuss how archaeology has given the past an exotic touch, as something fundamentally different to ourselves (Källén 2004: 22). Another influential concept that was introduced by Homi K. Bhabha is the notion of "the third space" (Bhabha 1994). The third space can be defined as an area where an encounter of two distinct (and unequal) social groups takes place, and where culture is disseminated and displaced from the interacting groups, making way for the invention of a hybrid identity. Within rock art research different aspects of the use of the 'Other' and the third space have been discussed by Lise Nordenborg Myhre (2004) and Johan Ling (2005). In an article from 2005 Ling discusses if the cultural meetings that took place at some of the maritime rock art concentrations on the West Coast could be interpreted as a kind of third space:

> "A space in between dominant social formations where cultural identity was being created, transmuted, articulated, communicated by the making, reading, interpretation and misinterpretation of rock art carvings" (Ling 2005: 455).

This definition of the third space could perhaps also count as a good description of the rock art sites that were being reused in different ways during the Early Iron Age, the difference being that the dialogue was now conducted between the living and the places and figures from the past (c.f Jones 2007: 3). It is also likely that people during the Iron Age could have regarded the figures as symbols made by the ancestors.

The disappearance of the south Scandinavian Bronze Age culture, as well as the adjoining tradition of making figurative rock art, has been explained as the effect of a number of combined factors. Kaul suggests that the disappearance of the rock art tradition was caused by a number of societal changes combined with, or caused by, massive changes in the societies north of the Alps. He also puts the disappearance in connection with possible religious changes at the end of the Late Bronze Age (Kaul 2005: 43). I would like to add another aspect to this by regarding the disappearance not only as a passive adjustment to external factors, but rather as a conscious act, performed by different groups on different occasions. For example, there are several rock art motifs from the West Coast of Sweden that can be dated to the Pre-Roman Iron Age while the tradition seems to have vanished in the Himmelstalund region by this time. By the conscious act of *not* making (figurative) rock art anymore the Early Iron Age societies in the Himmelstalund region dissociated themselves from the old tradition. But some of the figures – or the rocks themselves - continued to be important during the Iron Age, as shown by the reuse and revisits at rock art sites. This renegotiation of the meaning of the sites can also be viewed as a kind of dialogue with the past, probably understood by the Early Iron Age people as a dialogue with the ancestors. My point is that in times of cultural change places connected with the ancestors, such as cemeteries and rock art sites can become loaded with both positive and negative meanings. In other words - the past can become the 'Other'.

As we have seen, the dialogue with the past – and perhaps also with the ancestors - could be performed in many different ways. Some places were abandoned, some were continuously used and some were revisited. A few possible reasons for this have been proposed by Ulf Bertilsson, when it comes to the location of Iron Age graves at rock art sites:

> "First: the act expresses the striving for connection with old beliefs and cults. Second, it expresses respect for the ancestors and third, it expresses the continuous claim to adjacent territories" (Bertilsson 1987).

Besides these possible reasons for the reuse of rock art sites I would like to add one more possibility, and that is that the rock art sites could have played a problematic role, especially during the end of the Late Bronze Age and the Early Iron Age. Some of them were covered with thick layers of fire-cracked stones or soot and some of them were perhaps even destroyed. The graves or burnt mounds erected on top of some of the panels can also be viewed in this context, as a deliberate covering.

All of these possible approaches can be regarded as different dialogues with the past, dialogues that resulted in that a number of rock art sites continued to hold a meaningful, yet transformed meaning during the Iron Age. And it was through the dialogue with the past that this altered meaning was created.

Bibliography

Archives
The Royal Swedish Academy of Letters, History and Antiquities (Antikvarisk Topografiska Arkivet, ATA, Stockholm).

Literature

ARTELIUS, T.
2004 Minnesmakarnas verkstad: Om vikingatida bruk av äldre gravar och begravningsplatser. pp. 99-120. In Berggren, Å., Arvidsson, S. and Hållans, A-M. (eds.). *Minne och myt - konsten att skapa det förflutna*. Nordic Academic Press.

ARTELIUS, T. and LINDQVIST, M.
2007 *Döda minnen*. Stockholm: Riksantikvarieämbetet, Arkeologiska undersökningar. Skrifter 70.

BENGTSSON, L.
2004 *Bilder vid vatten. Kring hällristningar i Askum socken, Bohuslän*. Göteborg: Gotarc Serie C. Arkeologiska skrifter 51.

BENGTSSON, L. and LING, J.
2007 Scandinavia's most finds associated rock art sites. *Adoranten* 2007: 40–50.

BERTILSSON, U.
1987 *The rock carvings of northern Bohuslän. Spatial structures and social symbols*. Stockholm: Stockholm Studies in Archaeology No 7.

BHABHA, H. K.
1994 *The location of culture*. London and New York: Routledge.

BORNA-AHLKVIST, H.
2002 *Hällristarnas hem. Gårdsbebyggelse och struktur i Pryssgården under bronsålder*. Lund: Riksantikvarieämbetet. Arkeologiska undersökningar. Skrifter 42.

BORNA-AHLKVIST, H., LINDGREN-HERTZ, L., STÅLBOM, U.
1998 *Pryssgården - från stenålder till medeltid*. Arkeologisk slutundersökning RAÄ 166 och 167, Östra Eneby socken, Norrköpings kommun, Östergötland. Linköping: Riksantikvarieämbetet, Avd. för arkeologiska undersökningar. Rapport UV Linköping 1998:13.

BRADLEY, R.
2002 *The past in prehistoric societies*. London and New York. Routledge.

BROSTRÖM, S-G.
2007 *Himmelstalund. Del av RAÄ nr 1 i Östra Eneby socken, Östergötland*. BOTARK. Rapport 2007: 22.

BROSTRÖM, S-G. and IHRESTAM, K.
1993 Rapport över inventering och dokumentation av hällristningar utefter väg E4:as nya sträckning förbi Norrköping, sträckan Borgs kyrka – Åby. Stockholm. ATA. Unpublished report.

BURENHULT, G.
1973 *The rock carvings of Götaland (excluding Gothenburg County, Bohuslän and Dalsland). Part II*. Lund: Acta Archaeologica Lundensia. Series in 4. No 8.
1980 *Götalands hällristningar*. Stockholm: Del I. Theses and Papers in North-European Archaeology 10.

COLES, J.
2003 And on they went... processions in Scandinavian Bronze Age carvings. *Acta Archaeologica* 74, 211-250.

CORNELL, P. and FAHLANDER, F.
2002 *Social praktik och stumma monument. Introduktion till mikroarkeologi*. Inst. för Arkeologi, Göteborgs Universitet. Göteborg: Gotarc Serie C. Arkeologiska Skrifter No. 46.

ERICSSON, A. and NILSSON, P.
2007 *Hus och brunnar vid Himmelstalunds hällristningar*. Inför ny gång- och cykelväg. Arkeologisk förundersökning, del 2 och slutundersökning. Östra Eneby socken, Norrköpings kommun, Östergötland. Linköping: Riksantikvarieämbetet, UV Öst. Rapport 2007: 64.

FANON, F.
1962 *Jordens fördömda*. Göteborg.

FREDELL, Å.
2003 *Bildbroar. Figurativ bildkommunikation av ideologi och kosmologi under sydskandinavisk bronsålder och förromersk järnålder*. Göteborg: Gotarc Serie B. Gothenburg Archaeological Thesis no 25.

GERDIN, A-L. and MUNKENBERG, B-A.
2005 *Från mesolitisk tid till järnålder - Tanum, inte bara hällristningar*. Arkeologisk undersökning. Riksantikvarieämbetet, UV Väst. Västra Frölunda. Rapport 2005: 7.

GERRITSEN, F. A.
1999 The cultural biography of Iron Age houses and the long-term transformation of settlement patterns in the southern Netherlands. In Fabech, C. and Ringtved, J. (eds.). *Settlement and landscape. Proceedings of a conference in Århus, Denmark, May 4-7 1998*. Højbjerg. Jutland Archaeological Society.

GOLDHAHN, J.
2006 *Hällbildsstudier i norra Europa. Trender och tradition under det nya millenniet*. Institutionen för arkeologi, Göteborgs universitet. Göteborg: Gotarc Serie C. Arkeologiska skrifter No 64.

2007 Dödens hand – en essä om brons- och hällsmed. In Goldhahn, J. and Østigård, T (eds.). *Rituelle spesialister i bronse- og jernalderen.* Institutionen för arkeologi. Göteborgs Universitet. Göteborg: Gotarc Serie C. Arkeologiska skrifter No 65.

GUSTAV ADOLF, Prince. H K H (H. R. H.)
1904 Fornlämningar i trakten kring Norrköping. Linköping Utgifna av O. Klockhoff. *Meddelanden från Östergötlands fornminnesförening 1904, 1-6.*

HELANDER, A.
2005 *Kv Sällskapsdansen, St Johannes socken, Norrköpings kommun, Östergötland.* Linköping: Riksantikvarieämbetet, UV Öst. Rapport 2005: 47.

HOLTORF, C.
1999 Megaliths. Pp. 441–452. In Gustafsson, A. and Karlsson, H. (eds.) *Glyfer och arkeologiska rum – en vänbok till Jarl Nordbladh.* Göteborg: Gotarc Series A vol. 3.
2000-2007 *Monumental past: The life-histories of megalithic monuments in Mecklenburg-Vorpommern (Germany).* Electronic monograph. University of Toronto: Centre for Instructional Technology Development. http://hdl.handle.net/1807/245.
2008 The life-history approach to monuments: an obituary? In Goldhahn, J. (ed.). *Gropar and monument. En vänbok till Dag Widholm.* Kalmar: Kalmar Studies in Archaeology IV, 411 – 427.

JONES, A.
2007 *Memory and material culture.* Cambridge: Cambridge University Press.

KALIFF, A.
1997 *Grav och kultplats. Eskatologiska föreställningar under yngre bronsålder och äldre järnålder i Östergötland.* Uppsala: Inst. för arkeologi. AUN 24.
2007 *Fire, Water, Heaven and Earth: Ritual Practice and Cosmology in Ancient Scandinavia: An Indo-European Perspective.* Stockholm: Riksantikvarieämbetet.

KALIFF, A., BJÖRKHAGER, V., CARLSSON, T. and SKÖLDEBRAND, M.
1995 *Ringeby. En grav och kultplats från yngre bronsålder.* Linköping: Riksantikvarieämbetet, UV. Rapport 1995: 51.

KAUL, F.
2004 *Bronzealderens religion - studier af den nordiske bronzealders ikonografi.* København: Det Kongelige Nordiske Oldskriftselskab. Nordiske Forntidsminder Serie B, Bind 22.
2005 Arkæologiske undersøgelser ved helleristningerne. In Kaul, F., Stoltze, M., Nielsen, F. O. and Milstreu, G. (eds.). *Helleristninger. Billeder fra Bornholms bronzealder.* Rønne: Bornholms Museum/ Wormanianum: 134–40.
2006 Udgravninger ved helleristninger på Bornholm. *Adoranten* 2006: 50–63.

KRISTIANSEN, K. and LARSSON, T. B.
2005 *The rise of bronze age society. Travels, transmissions and transformations.* Cambridge: Cambridge University Press.

KÄLLÉN, A.
2004 *And through flows the river. Archaeology and the pasts of Lao Pako.* Uppsala: Institutionen för Afrikansk och jämförande arkeologi. Uppsala University. Studies in Global Archaeology 6.

LARSSON, L. K.
2005 Hills of the Ancestors. Death, Forging and Sacrifice on two Swedish Burial Sites. In Artelius, T. and Svanberg, F. (eds.). *Dealing with the dead. Archaeological perspectives on prehistoric Scandinavian burial ritual.* Riksantikvarieämbetet, Arkeologiska Undersökningar, Skrifter 65: 99-124.

LING, J.
2005 The fluidity of Rock Art. In Goldhahn, J. (ed.). *Mellan sten och järn. Del II. Rapport från det 9:e nordiska bronsålderssymposiet, Göteborg.* Göteborg: Gotarc Serie C. Arkeologiska skrifter No 59: 437–460.
2008 *Elevated Rock Art. Towards a maritime understanding of Rock Art in northern Bohuslän, Sweden.* Göteborg: Gotarc Serie B. Gothenburg Archaeological Thesis 49.

LUCAS, G.
2005 *The Archaeology of Time.* London: Routledge.

LUNDSTRÖM, P.
1965 *Gravfälten vid Fiskeby i Norrköping I. Studier kring ett totalundersökt komplex.* Stockholm: Kungl. Vitterhets Historie och Antikvitetsakademien.
1970 *Gravfälten vid Fiskeby i Norrköping II. Studier kring ett totalundersökt komplex.* Stockholm: Kungl. Vitterhets Historie och Antikvitetsakademien.

LØDØEN, T.
2006 Exploring the contemporary context of rock art. *Adoranten* 2006: 5–18.

MANDT, G.
2005 Kårstad i Stryn – møte mellom to kulttradisjonar. In Lødøen, T. and Mandt, G. *Bergkunst. Helleristningar i Noreg.* Oslo: Det Norske Samlaget.

MUNKENBERG, B.-A.
2004 Monumentet i Svarteborg. pp. 17–70. In Claesson, P. and Munkenberg, B-A. (eds.). *Gravar och ritualer.* Uddevalla: Bohusläns museum.

NILSSON, P.
2005a Om boplatslokalisering inom Bråbygdens hällristningsområden. In Goldhahn, J. (ed.). *Mellan sten och järn. Rapport från det 9:e nordiska bronsålderssymposiet.* Göteborg: Gotarc Serie C. Arkeologiska skrifter No 59: 419–435.
2005b *Fem hus från yngre bronsålder. Arkeologisk undersökning för fjärrvärmeledning vid Bråvallaområdet. Östra Eneby socken, Norrköpings kommun, Östergötland.* Linköping: Riksantikvarieämbetet, UV Öst. Rapport 2005: 59.
2007 *Fiskebyboplatsen. Arkeologisk förundersökning.* Linköping: Riksantikvarieämbetet, UV Öst. Rapport 2007: 28.
2008a *Härdar och nyupptäckta hällristningar vid Himmelstalund. Arkeologisk förundersökning/*

antikvarisk kontroll. Östra Eneby socken, Norrköpings kommun, Östergötland.Linköping: Riksantikvarieämbetet, UV Öst. Rapport 2008: 5.
2008b New discoveries of rock carvings and settlements at Himmelstalund. *Adoranten* 2007: 20–28.
2010a Reused rock art: Iron Age activities at Bronze Age rock art sites. In Fuglesvedt, I., Goldhahn, J. and Jones, A. (eds.) *Changing pictures: Rock art traditions and visions in northernmost Europe*. Oxford: Oxbow Books.
2010b A Life Aquatic? Looking at the relationships between settlements, rock-art and sea levels in the Himmelstalund region of eastern Sweden In Fredell, Å., Kristianssen, K. and Criado Boado, F. (eds.) *Representations and communications: Creating an archaeological matrix of late prehistoric rock art.*. Oxford: Oxbow Books.

NORDÉN, A.
1925 *Östergötlands bronsålder*. I beskrivande förteckning med avbildningar av lösa fynd i offentliga och enskilda samlingar, kända gravar samt hällristningar – översikt. Linköping: Henric Carlssons Bokhandels Förlag.
1936 Hällristningstraditionen och den urnordiska runskriften. Ett östgötskt runfynd i hällristningsmiljö och dess skrifthistoriska betydelse. In Thordeman, B. et al (eds.). *Arkeologiska studier tillägnade H.K.H. Kronprins Gustav Adolf*. Stockholm: Svenska Fornminnesföreningen.

NORDENBORG MYHRE, L.
2004 *Trialectic archaeology. Monuments and space in Southwest Norway 1700-500 BC*. Stavanger: AmS-Skrifter 18.

NORDSTRÖM, P.
1995 Arkeologiska undersökningar vid hällristningar. Unpublished undergraduate thesis in archaeology. Stockholm: Stockholm University.

PERSSON, H.
1998 *Himmelstalundsparken*. Norrköping (f d Östra Eneby socken), Östergötland. Arkeologisk förundersökning, april-juli 1994. Linköping: Östergötlands länsmuseum. Dnr: 326/93.

PETERSSON, M.
2006 *Djurhållning och betesdrift. Djur, människor och landskap i västra Östergötland under yngre bronsålder och äldre järnålder*. Uppsala: Inst. för arkeologi och antik historia, Uppsala universitet.

SAID, E. W.
1978 *Orientalism*. New York: Pantheon Books.

SELINGE, K-G.
1985 Om dokumentation av hällristningar. Metodiska synpunkter med östgötska exempel. *Fornvännen* 1985/2: 97–120.

STRÖMBERG, B.
2005 *Gravplats - gravfält: platser att skapa minnen vid - platser att minnas vid*. Institutionen för arkeologi, Göteborgs universitet. Göteborg: Gotarc Series B, Gothenburg archaeological theses, 35.

STÅLBOM, U.
1994 *Klinga. Ett gravfält: slutundersökning av ett gravfält och bebyggelselämningar från bronsålder och äldre järnålder*. Linköping: Riksantikvarieämbetet. Byrån för arkeologiska undersökningar. Rapport UV Linköping 1994: 11.
1998 Waste or what? Rubbish pits or ceremonial deposits at the Pryssgården site in the late Bronze Age. Lund: *Lund Archaeological Review* 1997: 21–35.

SØRENSEN, P. Ø.
2006 Arkæologiske udgravninger på Bornholm. *Adoranten* 2006: 64–73.

TEGNÉR, M.
2008 Nya gravar vid gamla högar. Bruket av det förflutnas platser vid järnålderns Öresund. In Carlie, A. (ed.). *Kulturella kontakter och samhällsutveckling i Skåne och på Själland under järnåldern*. Lund: Centrum för Danmarksstudier, Lunds Universitet.

THÄTE, E. S.
2007 *Monuments and minds. Monument re-use in Scandinavia in the second half of the first millenium AD*. Lund: Acta Archaeologica Lundensia Series in 4 No. 27.

TILLEY, C.
2008 *Body and image. Explorations in landscape phenomenology 2*. Walnut Creek: Left Coast Press.

WAHLGREN, K. H.
2002 *Bilder av betydelse: Hällristningar och bronsålderslandskap i nordöstra Östergötland*. Stockholm: Stockholm Studies in Archaeology 23.

WHITLEY, J.
2002 Too many ancestors. *Antiquity* 76: 119–126.

Recycled Memories:
The Past and Present in Early Iron Age Landscapes of Southern Germany

Matthew L. Murray

Abstract

Recycling is encountered at various scales in the mortuary landscapes of the Early Iron Age in southern Germany. In mortuary landscapes, acts of recycling included curation and redeposition of funerary materials, reuse of monuments for burial, and the citation of older monuments and mortuary features in new social discourses. These acts were intentional ways of enchaining objects, features, and individuals across time and space. They created theaters of "incorporated practices" that reinforced group identity and ideology, and through which the collective memory was transmitted or transformed. In this paper, I illustrate the idea of enchainment with examples of different scales of recycling.

At Tumulus 17 in the Hohmichele mound group, funerary remains from the primary burial event were curated and recycled within a single burial mound across several generations. Bettina Arnold (University of Wisconsin-Milwaukee) and I directed excavation of this mound in 1999-2000 as part of the "Landscape of Ancestors" project near the Heuneburg hillfort in southwestern Germany. Seriation of the primary and secondary graves in the mound and analysis of mound stratigraphy indicate that the monument was used intermittently as a cemetery from around 620 B.C. to about 400 B.C. Hearths and structured deposits in the mound fill reflect a long history of mound maintenance and visitation rituals that may be linked to ancestor veneration.

At a larger scale, mortuary monuments were integrated into complex structured landscapes that also incorporated linear earthworks and habitations. Monuments, as local theaters for incorporated practices, were linked to other places in choreographed landscapes where movement was manipulated to enhance the experience of certain monuments. I explore the process of landscape enchainment at the late Hallstatt-period Heuneburg as well as at the Early La Téne-period Glauberg in Hessen in west-central Germany.

Introduction: The Practice and the Scales of Recycling

> "Old things are always in good repute, present things in disfavor."
> - attributed to Publius Cornelius Tacitus (ca. 56-117)

> "To select well among old things, is almost equal to inventing new ones."
> - attributed to Nicholas Charles Trublet (1697-1770)

> "Sometimes old things need to go away. That way, we have room for the new things that come into our lives."
> - attributed to Randy K. Milholland (2005)

These aphorisms spanning nearly 2000 years of Western thought suggest that we – not just archaeologists but everyone – have ambiguous attitudes toward things from the past. In this volume, Dragos Gheorghiu and Phil Mason argue that recycling is a basic human process that involves the reuse of materials and ideas from the past. I propose that we can discern the recycling of materials, ideas, and identities at different scales in the Early Iron Age of Central Europe that enmeshed social actors in chains of relational processes explicitly linking the past and the present.

I use the idea of *scale* to separate human experience into nested spheres of interaction that involve increasingly more numerous and diverse actors, more and diverse opportunities for interaction, and greater integration of qualities of the built environment into theatres of social practice. Beginning with short-term events, such as funerals, I move to processes that created enduring elements of place, such as burial monuments, and then I explore the integration of such enduring elements into extensively engineered "memory-scapes."

The act and the process of recycling are encountered at various scales in the Early Iron Age. In mortuary landscapes, acts of recycling encompassed 1) curation and redeposition of funerary debris – both human and material, 2) reuse of monuments for burial, and 3) the citation of older monuments and mortuary features in new social discourses. These acts were intentional ways of enchaining objects, features, and individuals across time and space that created theatres of "incorporated practices." Such practices reinforced group identity and ideology, and transmitted or transformed collective memory. In this paper, I illustrate the idea of recycling as enchainment with examples of different scales of recycling.

Enchainment and Citation through Depositional Practices and Recycling

Early Iron Age archaeological landscapes are networks of deposits with human remains, objects, and fragments of objects. These elements are the material components of identity. John Chapman's (2000) theory of fragmentation and "enchainment" underscores how identity and social practice are constituted in a dynamic material environment of object, person, and place (see also Chapman and Gaydarska 2007; Jones 2007; Mills and Walker 2008). Enchainment is the process of creating links between people and place through time, using the materiality of social life. Material objects in place form local histories (or *biographies*) that influence future action. The identities of people are mutually constituted by the objects that they use and by the places that they inhabit, and the material environment is an active matrix for social reproduction (Chapman 2000: 4). Fragments of objects – or even human remains – are exchanged, linking places and people, and creating or maintaining a *chain* of relationships.

There is a permanent link between people and the distinct biographies of their material culture. According to Chapman (2000: 5): "Thus, people are exchanging themselves as they exchange polished stone axes and painted ceramics...." This idea resembles the Trobriand Island "Kula Ring" (Malinowski 1922), a network of life-long trading partners who circulated shell jewellery, fusing object biographies with personal histories. The distinction between enchainment and accumulation also resembles Mauss' (1954) distinction between "gift" and "commodity" exchange, where the gift represents enchained personal relations and commodity reflects impersonal concepts such as wealth and value (Chapman 2000: 48). According to Chapman (2008: 188), the process of enchainment organizes ("mobilizes") intersecting identities of things, persons, and places through "presencing" or the reference to elements that are absent, such as a deceased group member. Chapman interprets accumulation and deposition as a break in enchainment, but I suggest that deposits are nodes of object/person relationships that are physically "embedded" in the earth – essentially an "embodiment" of place. Accumulation through deposition creates links between objects and people that are incorporated into places, producing or reproducing personal and group identities. In a recent formulation of the idea of enchainment, Chapman (2008: 199) argues that the dispersion of fragmentary material culture across a landscape is an important social practice that establishes and maintains enchained social relationships across multiple scales.

Citation is a reference to something – or someone – in the past during materially constituted social practice (Mills and Walker 2008: 18). Citations of relationships develop and change through the enchainment of people and their materials, and alterations in enchainment (such as object use and deposition) reflect the restructuring of old relationships (Mills and Walker 2008: 18-19; Pauketat 2008). Citations to the past are the way that genealogies of practices (and of practitioners) are created, bridging the past and the present, and sometimes linking distant places through the combination of materials with different local origins (Mills and Walker 2008: 18). I will use a brief example from the Early Iron Age of southwest Germany to illustrate fragmentation, enchainment, and citation in action.

Tumulus 17: A "Chain" of Burned Bones and Bits of Pots

At Tumulus 17 in the Hohmichele mound group, funerary remains from the primary burial event were curated and recycled within a single burial mound across several generations. Bettina Arnold (University of Wisconsin-Milwaukee) and I directed excavation of this mound in 1999 and 2000 as part of the "Landscape of Ancestors" project (Fig. 1) near the Heuneburg hillfort in south-western Germany (Arnold 2002, 2003, 2004a, 2004b, 2005, 2008; Arnold et al. 2000, 2001, 2003; Arnold and Murray 2002). Seriation of the primary and secondary graves in the mound and analysis of mound stratigraphy reveal that the monument was used intermittently as a cemetery from around 600 B.C. to about 450 B.C. Hearths and structured deposits in the mound fill reflect a long history of mound maintenance and visitation rituals that may be linked to ancestor veneration (Schneider 2003).

Tumulus 17 was established in the early part of the sixth century BC, and it contained intact secondary inhumation burials that demonstrate that the mound was used into the late fifth century BC (Fig. 2). The founding funeral event was the burning of a body in place and its subsequent burial in the remains of the pyre. Although inhumation and cremation were distinct practices that must have reflected different ideas of the human body, in prehistory they both resulted in body residues that were "biographic aids to the remembering of past social circumstances" (David et al. 2008: 159). A small mound was then erected over the grave (Fig. 3). This mound surface was scattered with burned ceramic fragments, bronze debris, bone, and charcoal, which were left to erode toward the base of the mound. Ceramic refitting analyses undertaken by Seth Schneider (2003) determined that ceramic fragments in the mound deposit matched pottery remnants in the primary grave (Fig. 4). Residues from the founding funeral event evidently had been curated and subsequently deposited on the mound surface. This surface was then capped with a layer of sterile clay before it was reused as a burial place about 100 years later.

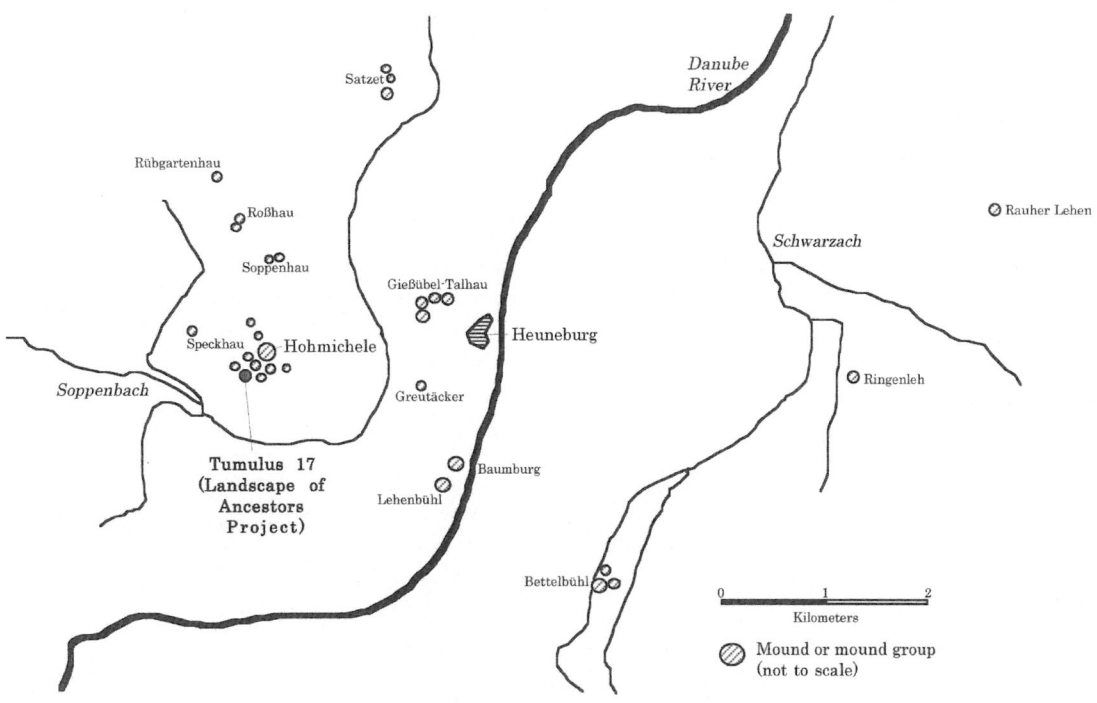

Fig. 1: *Location of Tumulus 17 (excavated in 1999-2000 as part of the Landscape of Ancestors project) in the Speckhau (Hohmichele) mound group near the Heuneburg. The map shows mounds and mound groups that are traditionally considered part of the Heuneburg mortuary landscape (adapted from Kurz, S. 2007, Figure 4).*

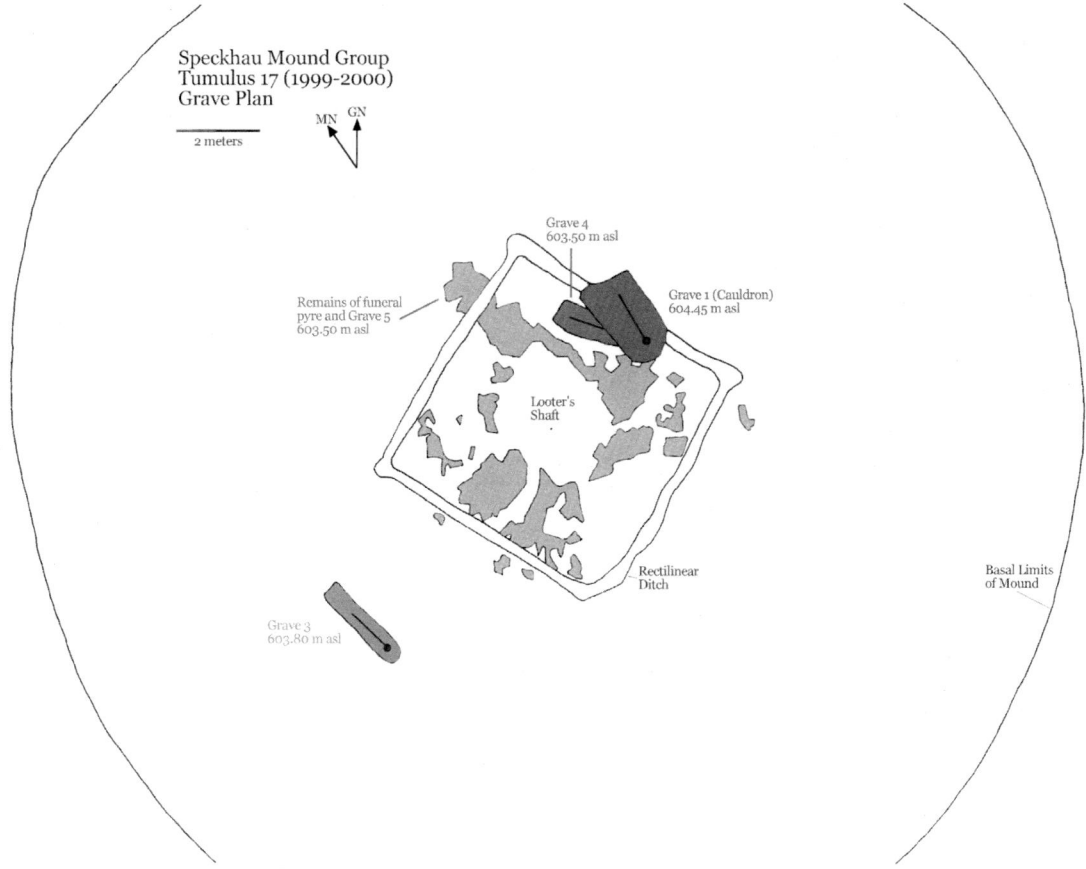

Fig. 2: *Plan view of Tumulus 17 showing the remains of the primary cremation grave (Gr. 5) in the central enclosure and the location and orientation of secondary graves.*

Fig. 3: *Photograph of the south and east profiles of the northwest quarter of Tumulus 17 in 1999 showing mound stratigraphy. The distinction between the original (inner) mound and the later (outer) mound layers can be clearly seen (author's photograph).*

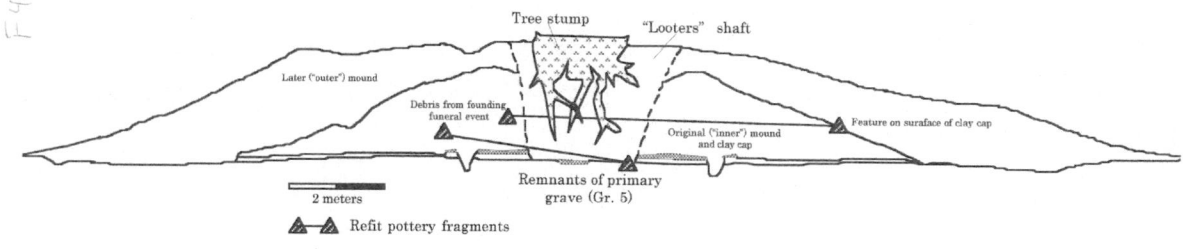

Fig. 4: *Idealized profile of Tumulus 17 showing original (inner) and later (outer) mound fill and the location of refit pottery fragments (pottery refit data from Schneider 2003: Figure 56).*

At the boundary between the original mound and later mound fill, we discovered charcoal deposits, burned areas, and small caches of stones and cultural debris. Schneider (2007) suggests that some of these features represent activities undertaken between mound phases when fire was used to "purify" the place and to prepare it for subsequent activity in the form of secondary interments. From one of these intact features, we recovered another pottery fragment that matches ceramic remnants in the primary grave (Schneider 2003). This fragment must have been placed in the mound several generations after the founding funeral event.

The primary grave at Tumulus 17 was looted and destroyed in antiquity. The mound was again explored in the late nineteenth century by an enthusiastic local forester, whose labours obscured traces of older looting. Schneider (2003, 2007) interpreted ceramic refits between the primary grave, inner mound surface, and a later non-mortuary feature to be evidence of curation for ancestor veneration rituals. I wonder if rather than *curation*, we have evidence of the *recycling* of funerary materials obtained by reopening the primary grave. The re-opening of mounds, when it can be dated, generally occurred within two generations, such as the paramount mounds at Grafenbühl and Magdalenenberg (Driehaus 1978).

We usually interpret the ancient disturbance of central burials in mounds like Tumulus 17 or the Hohmichele as grave robbery or desecration, either a crass economic practice or a more subtle social commentary (Driehaus 1978). But what if the intention was not to "rob" or "despoil" but to *recycle*? To collect the ancestors' parts and to re-link them in new chains – or perhaps, even to reconnect them to existing chains. Bones, fragments of burial pottery, and other funerary materials could be used to renew a relationship with the ancestors, to reinforce existing relationships by reference to past relationships, and to strengthen kinship groups by manipulation of the physical elements of their origin. Is grave disturbance evidence of repatriation and reburial? Were human remains that were recovered from cemeteries reburied or even reused in habitation contexts?

Tumulus 17 is a smaller version of the paramount Hohmichele mound located about 150 m away. Both mounds contained a disturbed primary grave and secondary burials. Like Tumulus 17, the Hohmichele contained evidence of hearths and structured deposits at the interface between mound construction and burial events. It also yielded evidence from the refitting of pottery, which indicates that older fragmented objects were recycled within the mound (Kurz and Schiek 2002; Schneider 2007).

The reuse of burial pottery at Tumulus 17 – and the Hohmichele – remind us that funerary rites may involve multiple events separated in time that link different kin aggregates, as described by Maurice Bloch (1971) for the Merina people of Madagascar. Among the Merina, the first funeral and physical interment of the body occurs soon after death and by necessity involves only local relatives and their supporters, since there is often little time to gather together more dispersed kin. A second event is organized years later, providing an opportunity for all kin group members to participate in a recovery of the deceased's body and its reburial. In a similar way, the links between pottery sherds at Tumulus 17 illustrate a process of fragmentation, "enchainment," and "citation" that sought to establish living links between social entities and their ancestors personified in a "corporeal" or embodied monument.

The Embodiment of Place at Tumulus 17

In previous papers and publications (Murray 1996a, 1996b, forthcoming), I have suggested that the "embodiment" of place is a form of "memory making" that links people and materials across space – and eventually through time – since many of these embodied places are long-lived and permanent (see also Bradley 1998; Tilley 1994, 2008; Lake 2007). This memory process was a community event, adapting the idea of "distributed memory" from Barth's (1987) work on cosmological practice. Ritual performances organize distributed memory and fix certain representations and meanings through group performances that subsequently inform perception and future performance. There are two general processes of bodily incorporation or "embodiment." First, in funerary practice, human remains are displayed and then deposited in the earth (literally, embodying the earth), sometimes in, or under, permanent monuments. In the second process, the landscape is "lived" or experienced through human body movement and sensation. Landscapes of the early Iron Age were intensively structured to "choreograph" these movements, often incorporating the "embodied" remains of ancestors. The choreography was a way of physically internalizing community memory of people and place. During the Early Iron Age, specific individuals were increasingly monumentalized and then referenced to reinforce particular relationships. This idea of embodiment and citation of place is related to John Barrett's (1999) concept of "inhabitation."

Monuments and Structured Landscapes: Theaters of Incorporated Practices

At a larger scale, mortuary monuments as embodied places were integrated into complex structured landscapes that also incorporated linear earthworks and habitations. In a recent formulation of the theory of enchainment, Chapman (2008: 199) argues that the dispersion of fragmentary material culture across a landscape is an important social practice that establishes and maintains enchained social relationships across multiple scales. I add to this the observation that embodied or inhabited places are related in similar chains of reference. Monuments, as local theatres for "incorporated practices," were linked to other places in "choreographed" landscapes where movement was manipulated to create links between features, to enhance experience of certain places, and to establish mnemonic patterns of social memories (Murray 1992, 1996a).

I will now explore landscape enchainment at the Late Hallstatt-period Heuneburg and the Early La Téne-period Glauberg in Germany. Both hillforts were part of a recent *Deutsche Forschungsgemeinschaft* project that examined the origin and development of early Celtic power centers (Krausse 2004, 2008; Krausse and Bofinger 2004), and the focus of new research beyond their existing walls reveals links between different contexts of recycling, enchainment, and citation.

The Heuneburg

The Heuneburg on the upper Danube River in southwest Germany is the archetypal elite fortified settlement of the Late Hallstatt period from the late seventh to the early fifth centuries BC (Kimmig 1983). Within 3000 meters of the hillfort, there are twelve known burial mound groups, including some of the largest monuments on the continent, such as the Hohmichele (Kurz and Schiek 2002). Several of these monuments have yielded the remains of Late Hallstatt and Early La Téne elite burials, including Tumulus 17, which I just discussed as an illustration of enchainment in place (Arnold and Murray 2002; Kurz and Schiek 2002).

Recent research at the Heuneburg has focused especially on the surroundings of the famous hillfort (Bofinger and Goldner-Bofinger 2008; Kurz, G. 2008; Kurz, S. 1998, 2000a, 2000b, 2007, 2008). This work has yielded evidence of monumental linear structures, including a system of intersecting ditch-and-wall systems (Fig. 5).

The systems encompassed various "outer" settlement areas that were roughly contemporary with the hillfort (Bofinger and Goldner-Bofinger 2008; Kurz, G. 2008; Kurz, S. 1998, 2000a), and they established links between the Heuneburg, outer settlements, and groups of elite burial monuments. In the core of the outer settlement, a massive stone "chambered" gate controlled movement across the site's interior (Kurz, G. 2008). People leaving the Heuneburg through this gate were oriented toward large burial mounds at Gießübel-Thalhau west of the hillfort. Additional monumental linear features also appear to link the hillfort and outer settlement with other large mounds to the south at Greutäcker in the "Südsiedlung" (Kurz, S. 1998) and Lehenbühl as well as a possible mound beneath the medieval fortification at the Baumburg.

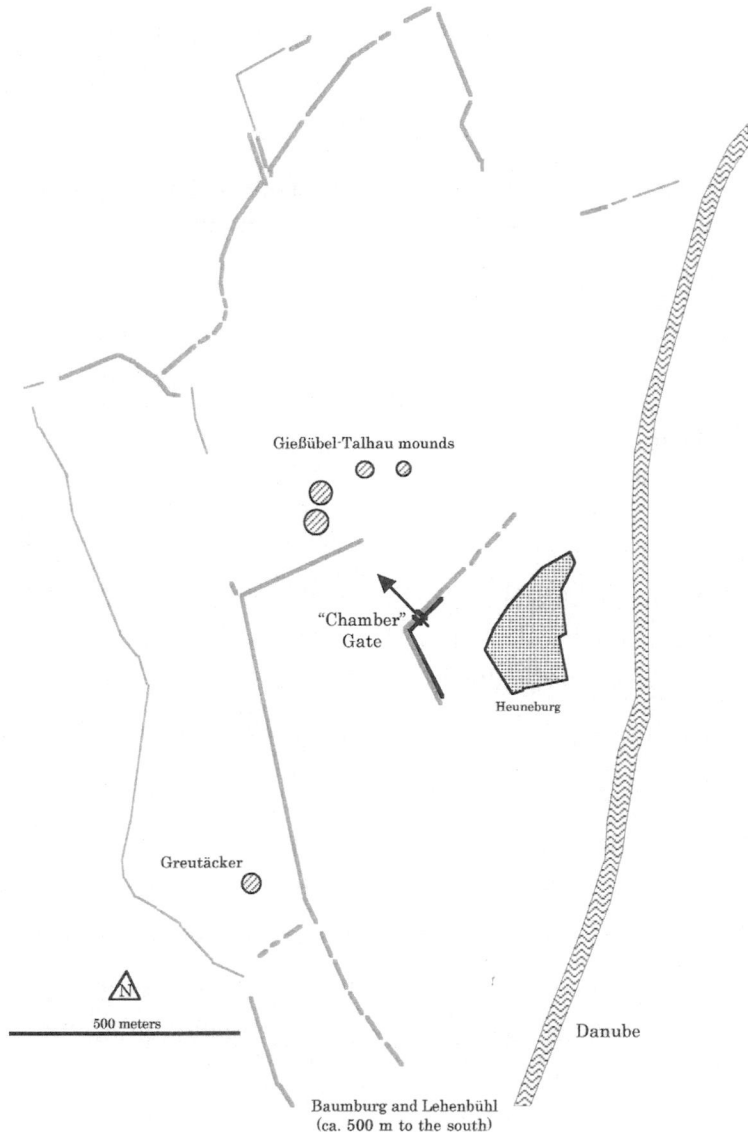

Fig. 5: *Plan of the structured landscape of the Heuneburg, including the hillfort and additional earthworks, as well as the "chamber" gate and its orientation toward the burial mounds at Gießübel-Thalhau. Ditches are shown in grey and wall remnants are shown in black; dashed lines indicate suspected earthworks (redrawn from Kurz, S. 1998; Kurz, S. 2008, Figure 1; and Kurz, G. 2008, Figure 10).*

At the Heuneburg, the hillfort, outer settlements, and mortuary monuments were once networked in an extensive complex of earthworks that regulated movement, choreographed performances, and guided experiences across the landscape. Some mortuary monuments at the Heuneburg, such as Tumulus 17, were re-used as burial places across multiple generations. Fragmentary materials from primary funeral events were recycled in these mounds as acts of enchainment that referenced deceased group members and created material and social links between generations. These places were then further referenced on a larger scale in a massive structuring of the Heuneburg landscape.

The Glauberg

Large-scale fieldwork at the Glauberg (Fig. 6) northeast of Frankfurt in west-central Germany has revealed portions of a highly complex structured landscape similar to the Heuneburg (Baitinger 2008; Baitinger and Pinsker 2002; Hansen and Pare 2008; Posluschny 2007). The Glauberg is a small basalt plateau that was fortified and occupied during the Late Hallstatt and Early La Tène periods in the sixth and fifth centuries BC. In 2001, three elite Early La Tène burials from the end of the fifth century BC were discovered at the foot of the hill (Herrmann 2002). The burials were originally marked by mounds that were integrated into of a system of ditches, walls, and enclosures around the hillfort (Hansen and Pare 2008). On the northwest side of Tumulus 1, a stone statue was found in a ditch along with fragments of two other anthropomorphic sculptures (Hermann 2002). The statue depicts a nearly life-size human figure bearing specific objects, in particular, a neck-ring and shield, which are strikingly similar to objects found in the central grave of Tumulus 1. It is likely that the statue originally crowned the mound and may have been intended to represent the individual beneath it.

The mound and statue, as an embodiment of personal identity, occupied a focal point in a monumental complex at the northern end of a 10-m-wide and 350-m-long earthen passageway (Hansen and Pare 2008). At the southern end of this passageway, the earthworks opened to embrace the plain, and ditches and associated walls continued to the east and west roughly perpendicular to the passageway. Interruptions in the earthworks indicate that they were not intended to be fortifications and they were constructed to encompass older landscape elements, such as several urn burials and a small group of mounds at Enzheimer Wald in the southwest. This process of citation linked older and younger mortuary elements. The system of linear features at the Glauberg effectively guided activity on the south side of the hillfort into the "passageway," thus focusing attention and action toward Mound 1 with its mortuary statue and its embodiment of a "heroic" identity. In this way, the built elements of the landscape choreographed cultural performances by directing and coordinating action across well-defined pathways toward corporeal monuments. The complex at the Glauberg likely incorporated knowledge from different systems of reference, for example, astronomical and cosmological observations also may have been incorporated directly into the system. Axel Posluschny (2007) suggests that the orientation of the passage leading to the elite mounds is aligned with a maximum lunar setting that occurs every 18 years (Southern Major Standstill), while the perpendicular ditches and walls reference the solstices.

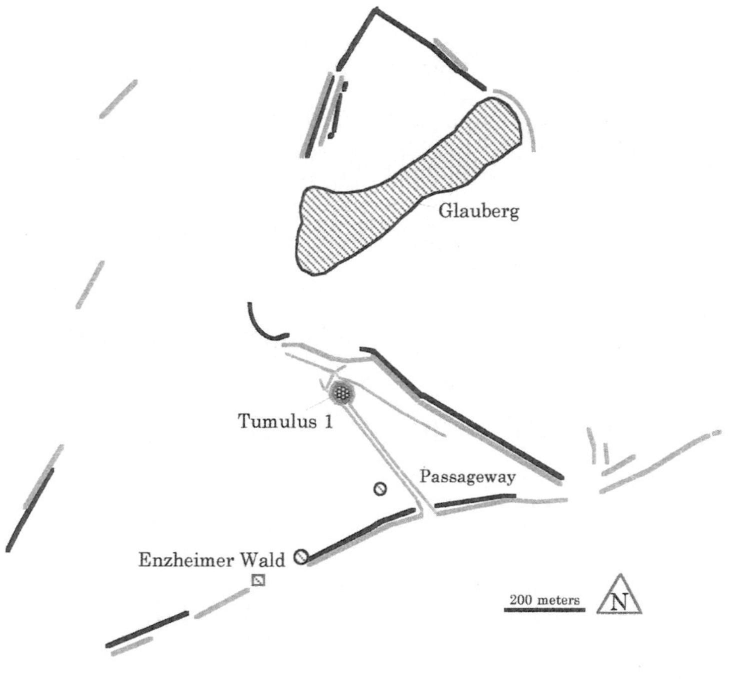

Fig. 6: *Plan of the structured landscape of the Glauberg showing the hillfort and a complex of earthworks, including the remains of Tumulus 1 at the northern end of a ditched passageway. The earthworks were interrupted to incorporate older mounds and urn graves at Enzheimer Wald. Ditches are shown in grey and embankments are shown in black (redrawn from Hansen and Pare 2008, Figure 1).*

Conclusion

Funeral, monument, landscape – in Early Iron Age Europe these are not just analytical scales of analysis but chains of increasingly complex relationships often based on the reuse or citation of old things and ideas. Funerals and the subsequent burial of funerary debris "embodied" places. In embodied places like Tumulus 17, the recycling of funerary fragments established or redefined links of identity and ideology across generations. Mortuary monuments were reused as burial places and were referenced in extensively structured landscapes where movement and experience were choreographed. Tumulus 17 also reveals that the mortuary landscape at the Heuneburg was a dynamic theatre of simultaneous and overlapping actions at different scales.

The despoiling of a grave must have been an act of potency and danger. In a 1978 article on grave robbery and desecration during the Early Iron Age, Jürgen Driehaus concluded that ultimately the issue is not why certain graves were targeted and others were not but why any grave was reopened at all. The idea of recycling may provide a partial answer – people were motivated to maintain the chains of relationships that defined their lives and identities by reconnecting themselves and their groups to parts of the past.

Bibliography

ARNOLD, B.
    2002 A Landscape of Ancestors: The Space and Place of Death in Iron Age West-Central Europe. In Silverman, H. and Small, D. B. (eds.) *The Space and Place of Death*, Washington D.C.: American Anthropological Association, pp. 129-143.
    2003 Landscapes of Ancestors: Early Iron Age Hillforts and their Mound Cemeteries. *Expedition: The Magazine of the University of Pennsylvania Museum of Archaeology and Anthropology* 45/1: 8-13.
    2004a Early Iron Age Mortuary Ritual in Southwest Germany: The Heuneburg and the Landscape of Ancestors Project. In Smejda, L. and Turek, J. (eds.) *Spatial Analysis of Funerary Areas*. Plzen: University of West Bohemia. pp. 148-158.
    2004b Early Iron Age Mortuary Ritual in Southwest Germany: The Heuneburg and the Landscape of Ancestors Project. In Smejda, L. and Turek, J. (eds.) *Spatial Analysis of Funerary Areas*, Plzen: University of West Bohemia. pp. 148-158.
    2005 Mobile Men, Sedentary Women? Material Culture as a Marker of Regional and Supra-Regional Interaction in Iron Age Europe. In Dobrzanska, H., Megaw, V., and Poleska, P. (eds.) *Celts on the Margin: Studies in European Cultural Interaction, 7th Century BC – 1st Century AD*. Krakow: Polish Academy of Sciences. pp. 17-26.
    2008 "Reading the Body": Geschlechtdifferenz im Totenritual der frühen Eisenzeit. In Kümmel, C. Schweizer, B., and Veit, U. (eds.) *Körperinszenierung – Objektsammlung – Monumentalisierung: Totenritual und Grabkult in frühen Gesellschaften*, Münster: Waxman. pp. 375-395.

ARNOLD, B. and MURRAY, M. L.
    2002 A Landscape of Ancestors in Southwest Germany. *Antiquity* 76:321-322.

ARNOLD, B., MURRAY, M. L., and SCHNEIDER, S. A.
    2000 Untersuchungen in einem hallstattzeitlichen Grabhügel der Hohmichelegruppe im 'Speckhau,' Markung Heiligkreuztal, Gemeinde Altheim, Landkreis Biberach. *Archäologische Ausgrabungen in Baden-Württemberg* 1999: 64-67.
    2001 Abschließende Untersuchungen in einem hallstattzeitlichen Grabhügel der Hohmichelegruppe im 'Speckhau,' Markung Heiligkreuztal, Gemeinde Altheim, Landkreis Biberach. *Archäologische Ausgrabungen in Baden-Württemberg* 2000: 67-70.
    2003 Untersuchungen in einem zweiten hallstattzeitlichen Grabhügel der Hohmichelegruppe im 'Speckhau,' Markung Heiligkreuztal, Gemeinde Altheim, Landkreis Biberach. *Archäologische Ausgrabungen in Baden-Württemberg* 2002: 78-81.

BAITINGER, H.
    2008 Der frühkeltische Fürstensitz auf dem Glauberg (Hessen). In Krausse, D. (ed.) *Frühe Zentralisierungs- und Urbanisierungsprozesse: Zur Genese und Entwicklung frühkeltischer Fürstensitze und ihres territorialen Umlandes*. Stuttgart: Konrad Theiss. pp. 39-56.

BAITINGER, H. and PINSKER, B.
    2002 *Glaube – Mythos – Wirklichkeit: Das Ratsel der Kelten vom Glauberg*. Stuttgart: Konrad Theiss.

BARRETT, J.
    1999 The mythical landscapes of the British Iron Age. In Ashore, W. and Knapp, A. B. (ed.) *Archaeologies of Landscape: Contemporary Perspectives*. Oxford: Blackwell. pp. 253-265.

BARTH, F.
    1987 *Cosmologies in the Making: A Generative Approach to Cultural Variation in Inner New Guinea*. Cambridge: Cambridge University Press.

BLOCH, M.
    1971 *Placing the Dead: Tombs, Ancestral Villages, and Kinship Organization in Madagascar*. London: Seminar Press.

BOFINGER, J. and GOLDNER-BOFINGER, A.
    2008 Terrassen und Gräben – Siedlungsstrukturen und Befestigungssysteme der Heuneburg-Vorburg. In Krausse, D. (ed.) *Frühe Zentralisierungs- und Urbanisierungsprozesse: Zur Genese und Entwicklung frühkeltischer Fürstensitze und ihres territorialen Umlandes*, Stuttgart: Konrad Theiss. pp. 209-228.

BRADLEY, R.
    1998 *The Significance of Monuments: On the Shaping of Human Experience in Neolithic and Bronze Age Europe*. London: Routledge.

CHAPMAN, J.
2000 *Fragmentation in Archaeology: People, Places and Broken Objects in the Prehistory of South Eastern Europe.* London: Routledge.
2008 Object Fragmentation and Past Landscapes. In Bruno, D. and Thomas, J. (eds.) *Handbook of Landscape Archaeology*, Walnut Creek, CA: Left Coast Press. pp. 187-201.

CHAPMAN, J. and GAYDARSKA, B.
2007 *Parts and Wholes: Fragmentation in Prehistoric Context.* Oxbow Books, Oxford.

DAVID, B., PIVORU, M., PIVORU, W., GREEN, M., BARKER, B., WEINER, J. F., SIMALA, D. KOKENTS, T., ARAHO, L., and DOP, J.
2008 Living Landscapes of the Dead: Archaeology of the Afterworld among the Rumu of Papua New Guinea. In Bruno, D. and Thomas, J. (eds.) *Handbook of Landscape Archaeology*, Walnut Creek, CA: Left Coast Press. pp. 158-166.

DRIEHAUS, J.
1978 Zum Grabraub in Mitteleuropa während der älteren Eisenzeit. In Jahnkuhn, H., Nehlsen, H. and Roth, Helmut (eds.) *Zum Grabfrevel in vor- und frühgeschichtlicher Zeit*, Göttingen: Vandenhoeck and Ruprecht. pp. 18-47.

HANSEN, L. and PARE, C.
2008 Der Glauberg in seinem mikro- und makroregionalen Kontext. In Krausse, D. (ed.) *Frühe Zentralisierungs- und Urbanisierungsprozesse: Zur Genese und Entwicklung frühkeltischer Fürstensitze und ihres territorialen Umlandes.* Stuttgart: Konrad Theiss. pp. 57-96.

Herrmann, F-R.
2002 Fürstensitz, Fürstengräber, und Heiligtum. In Baitinger, H. and Pinsker, B. (eds.) *Glaube - Mythos - Wirklichkeit: Das Ratsel der Kelten vom Glauberg*, Stuttgart: Konrad Theiss. pp. 90-107.

JONES, A.
2007 *Memory and Material Culture.* Cambridge: Cambridge University Pres.

KIMMIG, W.
1983 *Die Heuneburg an der oberen Donau.* Theiss, Stuttgart.

Krausse, D.
2004 Frühkeltische Fürstensitze: Ein neues Schwerpunktprogramm der Deutschen Forschungsgemeinschaft am Landesdenkmalamt Baden-Württemberg. *Denkmalpflege in Baden-Württemberg* 33/4: 237-245.

KRAUSSE, D. (ed.)
2008 *Frühe Zentralisierungs- und Urbanisierungsprozesse: Zur Genese und Entwicklung frühkeltischer Fürstensitze und ihres territorialen Umlandes.* Stuttgart: Konrad Theiss.

KRAUSSE, D. and BOFINGER, J.
2004 Neues DFG-Projekt: Genese und Entwicklung frühkeltischer Fürstensitze. *Archäologie in Deutschland* 2004 (4): 4.

KURZ, G.
2008 Ein Stadttor und Siedlung bei der Heuneburg (Gemeinde Herbertingen-Hundersingen, Kreis Sigmaringen). Zu den Grabungen in der Vorburg von 2000 bis 2006. In Krausse, D. (ed.) *Frühe Zentralisierungs- und Urbanisierungsprozesse: Zur Genese und Entwicklung frühkeltischer Fürstensitze und ihres territorialen Umlandes*, Stuttgart: Konrad Theiss. pp. 185-208.

KURZ, S.
1998 Neue Ausgrabungen im Vorfeld der Heuneburg bei Hundersingen an der oberen Donau. *Germania* 76: 527-547.
2000a *Die Heuneburg-Außensiedlung: Befunde und Funde.* Stuttgart: Konrad Theiss.
2000b Untersuchungen zur Herausbildung der hallstattzeitlichen Siedlung auf der Heuneburg. *Denkmalpflege in Baden-Württemberg* 29/1: 20-25.
2007 *Untersuchungen zur Entstehung der Heuneburg in der späten Hallstattzeit.* Stuttgart: Konrad Theiss Verlag.
2008 Neue Forschungen im Umfeld der Heuneburg. Zwischenbericht zum Stand des Projektes "Zentralort und Umland: Untersuchungen zur Struktur der Heuneburg-Außensiedlung und zum Verhältnis der Heuneburg zu umgebenden Höhensiedlungen." In Krausse, D. *Frühe Zentralisierungs- und Urbanisierungsprozesse: Zur Genese und Entwicklung frühkeltischer Fürstensitze und ihres territorialen Umlandes.* Stuttgart: Konrad Theiss. pp. 163-184.

KURZ, S. and SCHIEK, S.
2002 *Bestattungsplätze im Umfeld der Heuneburg.* Stuttgart: Konrad Theiss.

LAKE, M.
2007 Viewing Space. *World Archaeology* 39(1): 1-3.

MALINOWSKI, B.
1922 *Argonauts of the Western Pacific.* Dutton, New York.

MAUSS, M.
1954 *The Gift: Forms and Functions of Exchange in Archaic Societies.* Translated by Ian Cunnison. Glencoe, IL: Free Press.

MILLS, B. and WALKER, W. H.
2008 Memory, Materiality, and Depositional Practice. In Mills, B. and Walker, W. H. (eds.) *Memory Work: Archaeologies of Material Practices*, Santa Fe: School for Advance Research. pp. 3-23.

MURRAY, M. L.
1992 The Archaeology of Mystification: Ideology, Dominance, and Urnfields of Southern Germany. In Goldsmith, A. S., Varvie, S. Selin, D. and Smith, J. *Ancient Images, Ancient Thought: The Archaeology of Ideology.* Calgary: University of Calgary. pp. 97-104.
1996a Socio-political Complexity in Iron Age Temperate Europe: A Dialectical Landscape Approach. In Meyer, D. A., Dawson, P. C., and Hanna,

D. T. (eds.) *Debating Complexity*, Calgary: University of Calgary. pp. 406-414.

1996b Viereckschanzen and Feasting: Socio-Political Ritual in Iron Age Central Europe. *Journal of European Archaeology* 3/2 (1995): 125-151.

(forthcoming) Landscapes of Ancestors? The Structuring of Space around Iron Age Funerary Monuments in Central Europe. In Hill, E. and Hageman, J. (eds.) *The Archaeology of Ancestors*.

PAUKETAT, T. R.

2008 Founders' Cults and the Archaeology of *Wakan-da*. In Mills, B. and Walker, W. H. (eds.) *Memory Work: Archaeologies of Material Practices*. Santa Fe: School for Advance Research. pp. 61-79.

POSLUSCHNY, A. G.

2007 From Landscape Archaeology to Social Archaeology: Finding Patterns to Explain the Development of Early Celtic "Princely Sites" in Middle Europe. In Clarke, J. T. and Hagemeister, E. (eds.) *Digital Discovery: Exploring New Frontiers in Human Heritage*. Budapest: Archaeolingua. pp. 117-127.

SCHNEIDER, S. A.

2003 *Ancestor Veneration and Ceramic Curation: An Analysis From Speckhau Tumulus 17, Southwest Germany*. Unpublished MA thesis, Milwaukee: University of Wisconsin.

2007 Ashes to Ashes: The Instrumental Use of Fire in West-Central European Early Iron Age Mortuary Ritual. In Gheorghiu, D. *Fire as an Instrument: The Archaeology of Pyrotechnologies*, BAR International Series, Oxford: Archaeopress,. pp. 85-95.

TILLEY, C.

1994 *A Phenomenology of Landscape: Places, Paths and Monuments*. Oxford: Berg.

2008 Phenomenological Approaches to Landscape Archaeology. In David, B. and Thomas, J. (eds.) *Handbook of Landscape Archaeology*. Walnut Creek, CA: Left Coast Press, pp. 271-284.

Ancestral Places:
The Creation and Recycling of Monumental Landscapes in South-Eastern Slovenia in the 1st Millennium BC and the 1st Millennium AD

Phil Mason

Abstract

The reuse of monumental landscapes in later periods is a familiar feature in many areas in Europe. The monumental landscapes of south-eastern Slovenia were largely a creation of the Early Iron Age. The paper seeks to show how Middle and Late Bronze habitation and mortuary sites were monumentalised by the construction of barrows and hillforts, resulting in the creation of highly visible places in the landscape. Where such earlier places were not present, earlier material was often incorporated into monuments to link such places to the past. Such monumentalised places were often incorporated into the landscapes of the Late Iron Age, the Roman period and Early Medieval period.

It is posited here that this incorporation of earlier elements in a new whole was a means of legitimating a changing socio-political milieu through the recycling and, as such, reinterpretation of real and fictitious ancestral places, which often lay in dominant, highly visible positions in the landscape.

Introduction

The reuse of earlier monumental landscapes in later periods is a familiar feature in many areas in Europe. This paper seeks to examine how the monumental landscapes of the Early Iron Age in south-eastern Slovenia were created, incorporating and transforming many elements of the preceding Middle Bronze Age and Late Bronze Age landscape to create highly visible places. It goes on to examine how these places were subject to continuous use and reinterpretation both in the Early Iron Age and in later periods.

The area under consideration corresponds to the modern regions Dolenjska and Bela krajina between the rivers Sava and Kolpa in South-eastern Slovenia. It is a landscape of broad river valleys, gorges and lowland basins, often sharply divided from one another by heavily forested hilly, karsitified interfluves. The interfluves also contain discrete upland basins, which are visually dominated by the surrounding plateau. The distant Alps are visible on a clear day throughout Dolenjska and from the crest of the Gorjanci hills. This range bounds the northern horizon of Bela krajina, making it the only part of Slovenia, where the Alps are not visible. Here the western horizon is bounded by the high Dinaric hills of Kočevski rog, whilst the southern horizon is dominated by the summits of the mountains in the Gorski Kotar region, which rise above the Gulf of Kvarner. The position of the region makes it a major corridor between the Po plain, the Eastern Alps, the Pannonian basin and the Balkans (Dular and Tecco Hvala 2007: 44-63, 217-223; Mason 1996a: 1-8).

The pre-Early Iron Age Landscape (Fig. 1)

The earliest evidence of monumentally in central and south-eastern Slovenia is found with the appearance of defended Copper Age settlements in the late 4th and early 3rd millennium B.C (Dular 2001: 89-106; Mason and Andrič 2009: 332). The succeeding Early Bronze Age and Middle Bronze Age periods appear to have been characterised by small dispersed farmsteads with isolated cremation or inhumation graves (Dular 1999: 92-95; Horvat 2005: 183-184; Mason 2009: 217-219). There is no evidence for defended settlements or barrows in Dolenjska and Bela krajina in these periods, a sharp contrast to the Early and Middle Bronze Age in Central Europe, or in the western and southern Balkans (Coles and Harding 1979: 40-45, 198-199).

The Late Bronze Age landscape was dominated by extensive open settlements on river terraces from the 11th to 9th centuries BC (Teržan 1999: 102-104, 107; Mason 2001b: 24; Mason 2006: 230-231). Upland settlements on the edges of interfluves above river valleys or on major route ways began to appear in the 10th century BC and increase in numbers in the 9th and 8th centuries (Dular and Tecco Hvala 2007: 132-135; Mason 2008: 97; 2009: 228-230; Teržan 1999: 111-119). They occupy highly visible positions, but are relatively small and are largely defended by palisades or rare wooden box ramparts. The mortuary record is characterised by flat cremation cemeteries without much evidence of social differentiation, although the presence of formal burial areas in itself is an important departure from previous periods.

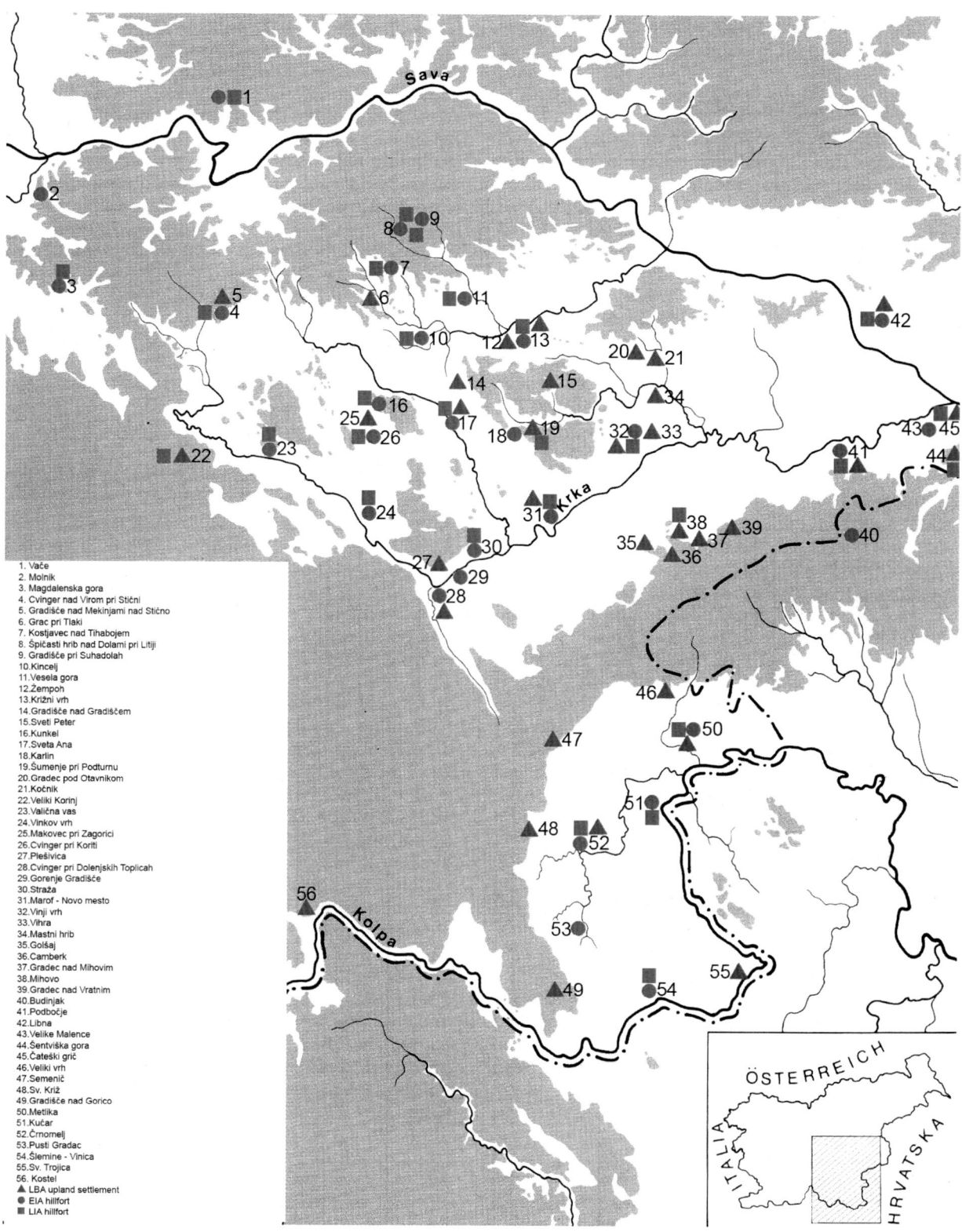

Fig. 1: *Dolenjska and Bela kraijina in the 1st millennium BC, showing major settlements (After Dular 1993: 103, fig. 1; with additions from Dular 1985: 31, fig. 12; ANSl 1975; Dular and Tecco Hvala 2007; drawn by Ildikó Pintér). 1- Vače; 2- Molnik; 3- Magdalenska gora; 4- Cvinger nad Virom pri Stični; 5- Gradišče nad Mekinjami nad Stično; 6- Grac pri Tlaki; 7- Kostjavec nad Tihabojem 8- Špičasti hrib nad Dolami pri Litiji; 9- Gradišče pri Suhadolah; 10- Kincelj; 11- Vesela gora; 12- Žempoh; 13- Križni vrh; 14- Gradišče nad Gradiščem; 15-Sveti Peter; 16- Kunkel; 17- Sveta Ana; 18- Karlin; 19- Šumenje pri Podturnu; 20- Gradec pod Otavnikom; 21- Kočnik; 22- Veliki Korinj; 23- Valična vas; 24- Vinkov vrh; 25- Makovec pri Zagorici; 26- Cvinger pri Koriti; 27- Plešivica; 28- Cvinger pri Dolenjskih Toplicah; 29- Gorenje Gradišče; 30- Straža; 31- Marof – Novo mesto; 32- Vinji vrh; 33- Vihra; 34- Mastni hrib; 35- Golšaj; 36- Camberk; 37- Gradac nad Mihovim; 38- Mihovo; 39- Gradec nad Vratnim; 40- Budinjak; 41- Podbočje; 42- Libna; 43- Velike Malence; 44- Šentviška gora; 45- Čateški grič; 46- Veliki vrh; 47- Semenič; 48- Sv. Križ; 49- Gradišče nad Gorico; 50- Metlika; 51- Kučar; 52- Črnomelj; 53- Pusti Gradac; 54- Šlemine – Vinica; 55- Sv. Trojica; 56- Kostel.*

Early Iron Age Monumental Landscapes

This situation was transformed at the beginning of the Early Iron Age. The landscape in central Slovenia was now dominated by monumental structures - hillforts and barrow groups that bear witness to the power of the Early Iron Age elites and their connections with communities in the wider North Adriatic, Alpine and Pannonian worlds (Mason 1996a; 1996b: 273-282; 2008: 97-106). Inspiration for barrows can be found in the Middle Bronze Age past and Late Bronze Age/ Early Iron Age present of the surrounding regions, but not within the region itself, whilst models for hillforts existed in the distant past of south-eastern Slovenia, as well as in the recent past and present of the surrounding regions (Coles and Harding 1979: 40-45, 198-199, 339-352, 362-365, 444-445; Dular 2001: 89-106).

Early Iron Age Hillforts (Fig. 1)

Early Iron Age hillforts appeared in the 8th century BC. They are fewer in number than the upland sites of the preceding period, but 16 of the 24 known sites have evidence of Late Bronze Age activity or occupation. The Early Iron Age hillforts occupy similar positions to those of the preceding Late Bronze Age upland sites on prominent locations in the landscape. However, they are much larger, ranging from 3 hectares at 22 hectares in area (Mason 2008: 97). Only at Cvinger is there direct evidence of the monumentalisation of a Late Bronze Age wooden rampart by a stone rampart enclosing the same area in the Early Iron Age (Dular and Križ 2004: 207-250). Other sites that exhibit continuity, such as Vinji vrh, probably expanded in the Early Iron Age, so that the earlier Late Bronze Age occupation is to be expected in a more limited area within the later hillfort (Fig. 2).

The Early Iron Age hillforts are generally univallate, being surrounded by massive ramparts (Guštin 1978: 100-121), which were important symbols of the power of the elite and/or the communities, resident within, or dependent on them. The ramparts were revetted on the interior and exterior with massive drystone walls, the rubble and earth rampart fill containing occupation debris (animal bone and potsherds) and large amounts of iron slag. The stone used in rampart construction was probably derived from field clearance and could thus symbolise the agricultural land associated with the hillfort, the occupation debris in the rampart fill the community itself (Mason 1996b: 276-277; 2008: 97, 104). Thus, the recycling of this material connected the living community to the ramparts that encircled it, which themselves were symbolic of the wider agricultural land available to the community resident within.

The presence of metalworking debris in hillfort ramparts along with settlement debris, which symbolically reinforces the hillfort rampart as a liminal zone. In this case, it bounds the living community, the symbolism here potentially being connected with regeneration and fertility. There is also a case where a layer containing burnt fine ware was deliberately incorporated as a foundation deposit beneath a new Early Iron Age rampart at Libna (Novaković, pers. comm.). This may well commemorate communal feasting, perhaps immediately prior to rampart construction, but it is important that the remains of

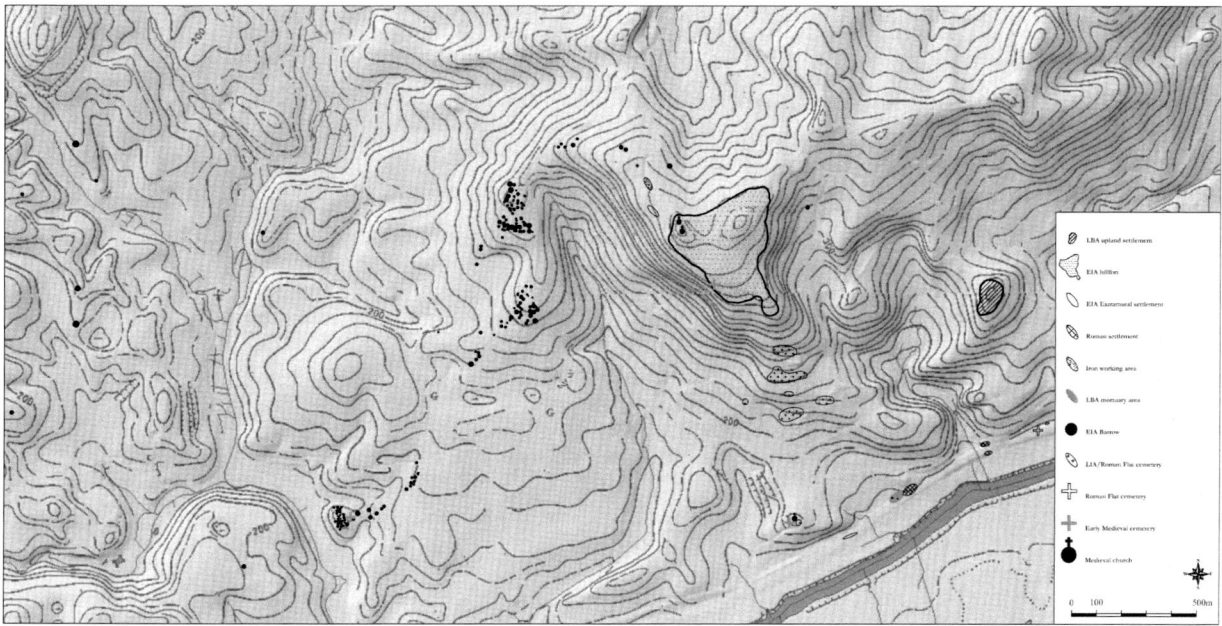

Fig. 2: *Plan of the late prehistoric settlement complex at Vinji vrh (Source: Agencija RS za okolje; data adapted from A. Dular 1991, fig. 3; with addition of recent data; drawn by Ildikó Pintér).*

this communal event were deliberately transformed by fire before incorporation in the rampart. Thus the monumental rampart of the hillfort served as a marker of the community within or dependant on it, the elite connected with it, the land and resources available to the community and elite and their and by inference its contacts and competition with the wider world.

Barrows

Earthen barrows, sometimes with central stone chambers, are also a new defining monumental element in the Early Iron Age landscape of south-eastern Slovenia, as well as in the wider Eastern Alpine region and Pannonia in the 8th century BC. These usually cluster around the hillforts. Cremations graves were replaced by inhumation graves in the 7th century BC, and a new form of burial arrangement appeared, the so-called "Sippenhügel" form, with concentric rings of inhumations within the barrow, without central elite burials. These are interpreted as the burial places of corporate groups, associated with the hillforts centres (Dular and Tecco Hvala 2007: 237-238; Mason 1996a: 20, 76-85; 2008: 99).

The concentrations of barrow groups around hillforts may be considerable. Thus, the hillfort at Vinji vrh is surrounded by c. 150 barrows (Fig. 2), as is the major hillfort centre at Stična, but the number of barrows is often smaller. The barrow groups create a monumental approach to the hillfort centre itself, strung out along the approach to the entrance at Vinji vrh (Fig. 2), or clustering like a facade before the entrances at Kučar (Fig. 3), Cvinger, and Vinkov vrh (A. Dular 1991: fig. 3; J. Dular, Ciglenečki and A. Dular 1995: 8-9, fig. 2; Mason 2008: 99)

Some of these barrow groups, like the hillforts themselves, are associated with earlier, Late Bronze Age mortuary areas. The clearest case is that of the Kapiteljska njiva barrow cemetery, which overlies an earlier, Late Bronze Age flat cemetery. It is not clear if the intention here is to incorporate an earlier, ancestral place, associated with the later burial population or to violently annex it. Incorporation of an earlier ancestral burial place is clearer at the Kučar hillfort centre, where the large southern barrows are closely associated with a Late Bronze Age cemetery (Mason 2008: 103, 105). Similar associations can also be seen at Molnik (Dular and Tecco Hvala 2007: 160-161).

Isolated Barrows

The large number of isolated barrows and barrow groups outside the narrow area of the hillfort centre may well mark formalised lines of travel through the landscape, but in some cases they are also connected with Late Bronze Age settlement sites. The clearest example of such an association is at Grofove njive, where an isolated barrow is located beside a path that

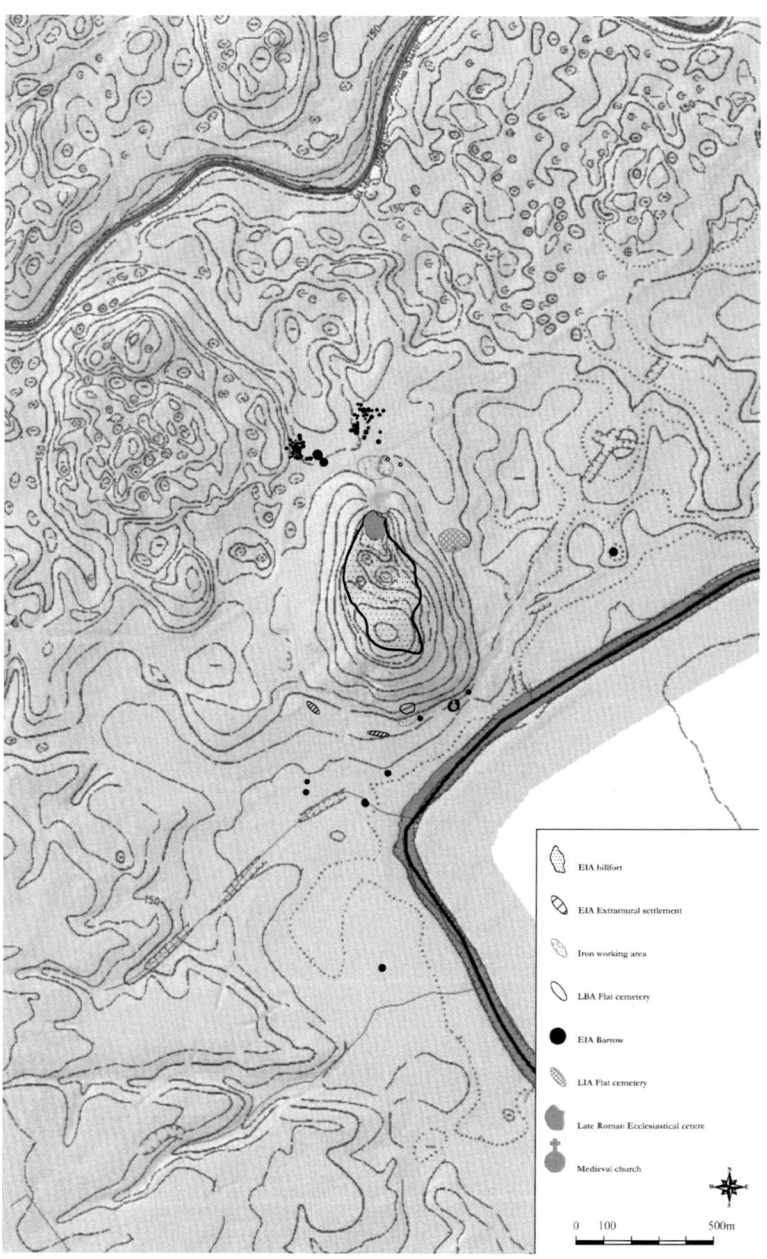

Fig. 3: *Plan of the late prehistoric settlement complex at Kučar pri Podzemlju (Source: Agencija RS za okolje; data adapted from J. Dular, Ciglenečki and A. Dular 1995: 8, fig. 2; with addition of recent data; drawn by Ildikó Pintér).*

runs north towards the Libna hillfort, at the point where it runs directly past the edge of a large Late Bronze Age settlement that is at least 300 years earlier (J. Dular and Tecco Hvala 2007: 294; Mason 2006: 230-231). The isolated barrow at Gomilca near Griblje is similarly placed on an earlier, Late Bronze Age settlement (Dular 1985: 74; Mason 2001b: 24). Thus these barrows mark a potentially fictitious ancestral place in the landscape, or annexe a pre-existing ancestral place to the existing elite by its association with their immediate ancestors.

Isolated barrows may also be directly connected with earlier mortuary sites, such as at Dolge njive, south of the Vinji vrh hillfort, where early Iron Age barrows were placed directly on three Late Bronze Age stone burial platforms (Mason 2004: 123-125) (Fig. 2).

Some of the isolated barrows that are associated with pathways do not have clear associations with earlier sites. This is the case with the barrows at Otočec that mark a route descending to the river Krka and continuing south along a route marked by the Ratež-Brusnice barrow group (Dular and Tecco Hvala 2007: 323, 331, 333, fig. 250, fig. 251; Križ 1989: 213-214; Teržan 1974: 33-66). The single excavated barrow at Otočec was constructed around a central core of organic rich soil, possibly derived from settlement debris. This may be an attempt to symbolically link the isolated barrow to a settlement. Similar activity may be represented in the presence of lithic material in the ploughed-out barrows at Kapiteljska njiva and at Kozjane (Križ pers. comm.; Mason 2001a: 24).

However, the lack of earlier settlement activity in the immediate vicinity of these sites may be more apparent than real. This is borne out by recent fieldwork in the area of Mačkovec to the west of Novo mesto, where a route descends from the upland rim to the Krka valley. The route was marked by two barrows, at the northern end and western side of the central part of a dry valley (Mason and Britovšek 2007; 2008: 72-73, 106-107, 116-120, 136; Udovč 2009: 6). Both barrows date to the 6th and first half of the 5th century BC, dates, which match those of other excavated barrows, such as that at Otočec, Brusnice and Grofove njive. More importantly, however, there is strong evidence of Middle Bronze Age settlement in the area, close to and overlooked by the barrows.

Continuity and Change in the Late Iron Age Period (Fig. 1)

Barrow burial ceased and the Early Iron Age elites disappeared at the beginning of the Late Iron Age in the late 4th and early 3rd centuries B.C, but the monumental landscape of hillforts and barrows continued in use. Many of the hillforts continued as centres of power and were elaborated with new ramparts in the 2nd century BC. The position of the barrow cemeteries is more ambivalent. The Middle La Tène flat cemetery is deliberately placed in opposition to the barrows at Kapiteljska njiva. The large Late La Tène flat cemeteries seem to be consciously sited to avoid earlier barrow cemeteries, whilst maintaining a close association with the hillforts themselves, as is the case with Strmec near Vinji vrh (Fig. 2), Zemelj-Jurajevčičeva njiva near Kučar (Fig. 3) and Beletov vrt near Novo mesto (Dular and Tecco Hvala 2007: 177-179; J. Dular, Ciglenečki and A. Dular 1995: 9; Božič 2008: 13-28). It would thus appear that there was a tendency by some groups to avoid direct association with the barrows that may have been seen as representative of the previous Early Iron Age elite, whilst maintaining an association with the hillfort centres. However, there is some evidence of secondary Late Iron Age burials inserted into barrow mounds from the isolated barrow at Medvedjek, in the Brodaričeva loza barrow group near Kučar (Fig. 3), as well as in the Laščik and Preloge barrow groups at Magdalenska gora, (Breščak 1990: 43-44; J. Dular, Ciglenečki and A. Dular 1995: 9; Tecco Hvala et al. 2004: 107, 109), which suggests that other groups or individuals actively sought association with these barrows, or rather with what these barrows represented. It is also clear that even if there was no apparent direct use of barrows, they were respected as elements in the landscape, a fact which was more much more apparent in the succeeding Roman period. Thus the isolated barrow that later formed the focus of a Roman cemetery at Mačkovec was surrounded by tree-throw hollows, suggesting that it was probably respected by cultivation (Plate 1).

Recycling the Iron Age Monumental Landscape in the Roman Period (Fig. 4)

Re-use and reinterpretation of the Early Iron Age monumental landscape becomes more apparent in the Early Roman period in south-eastern Slovenia. Many Early Iron Age hillforts were occupied in the final Iron Age, but seem to have been abandoned after this (e.g., Vinji vrh, Metlika, Črnomelj, Novo mesto). There is evidence for a shift to an adjacent location in some areas. Thus the Late Iron Age settlement in the town centre of Črnomelj was succeeded by a Roman small town or villa at on the adjacent Okljuk meander in Loka (Mason 1998: 285), whilst the Marof enclosure and the Late Iron Age settlement on the Kapitelj hill in the historic town centre of Novo mesto was probably succeeded by the large villa at Groblje (ANSl 1975: 226). However it should be noted that there is some evidence for Roman activity on the Kapitelj hill (Breščak pers. comm.)

However, the lack of extensive excavation within hillforts makes it impossible to categorically exclude the possibility of continuing occupation on a limited scale, or the presence of ritual structures or activity of the type that can be observed in other parts of the

Plate 1: *The Roman flat cremation cemetery at Mačkovec (Photo: Marko Pršina; Archive ZVKDS, CPA)*

western provinces, where Roman temples were often built within hillforts, e.g. at Maiden Castle in southern Britain (Sharples 1991: 16, 130). It is also noteworthy that burial continued in the large cemeteries associated with the Iron Age hillforts at Mihovo, Novo mesto and Vinji vrh (Fig. 2). Thus part of the population clearly identified with the pre-existing late pre-Roman Iron Age centres, which evidently maintained a symbolic meaning within the changing landscape of the early 1st millennium AD.

The changing landscape is most clearly evidenced by the appearance of a large number of relatively small rural flat cremation cemeteries, many of which exhibit associations with Early Iron Age barrows. The cemeteries at Mačkovec (plate 1), Medvedjek, Velika Dobrava, and possibly Smolenja vas, Hrast pri Jugorje – Grive and Griblje – Kohane, utilise isolated Early Iron Age barrows as focii, whilst for the isolated Early Iron Age barrow at Otočec has a single Roman period grave inserted into the mound. Other cemeteries are associated with Early Iron Age barrow groups close to major Early Iron Age centres, such as at Dobrnič or Late Bronze Age cremation cemeteries, as at Metlika – Borštek (Mason 2012, 390-399).

The foundation of new cemeteries at the beginning of the Roman period may represent a strategy employed by local groups to identify with real or fictitious ancestral places in the landscape, places which were not tainted by association with the immediate pre-Roman socio-political centres, whose cemeteries continued in use. Such places were potentially employed to exercise claims to control over land in the context of the changing socio-political situation of the beginning of the Roman occupation. The use of these rural cemeteries spans the period from the 1st century AD to the 3rd century AD, when changing political and social conditions resulted in their gradual disuse and abandonment (Mason 2012, 399-400).

Recycling the Iron Age Monumental Landscape in the Early Medieval and High Medieval Period (Fig. 4)

The Late Roman period saw the reuse of many hillfort sites in Dolenjska and Bela krajina, as the site of ecclesiastical complexes, e.g. at Kučar (J. Dular, Ciglenečki and A. Dular 1995: 71-190) (Fig. 3) and Gradac nad Mihovim (Dular 2008: 126-129) (for further examples see: Ciglenečki 1987). Such sites could have functioned as refuge forts for the surrounding population and possibly as administrative centres in the period between the mid 4th century AD and end of the 6th century AD, when central civil authority declined. It is possible that the poorly preserved 6th century cemetery on the south-western edge of the Vinji vrh

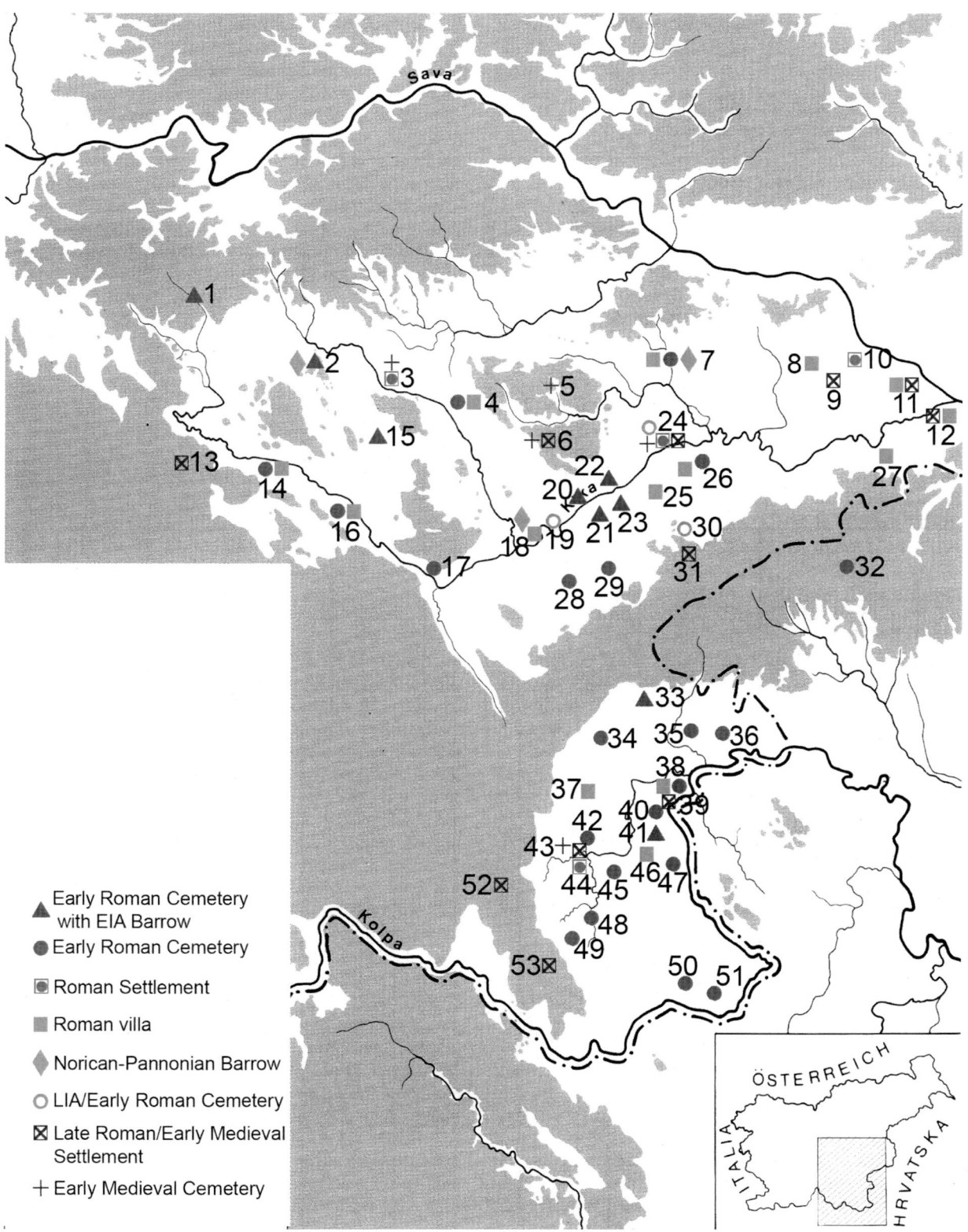

Fig. 4: *Dolenjska and Bela kraijina in the 1st millennium AD, showing Roman and Early Medieval settlement and mortuary sites, including those with evidence of re-use of later prehistoric sites (After Dular 1993: 103, fig. 1; with additions from Dular 1985: 31, fig. 12; ANSl 1975; Dular and Tecco Hvala 2007; drawn by Ildikó Pintér).* 1- Velika Dobrava; 2- Medvedjek; 3- Trebnje; 4- Jezero; 5- Sveti Peter; 6- Šumenje pri Podturnu; 7- Zloganje pri Škocjanu; 8- Veliki Dol pri Veliki Vasi; 9- Velike njive pri Veliki Vasi; 10- Drnovo; 11- Pečina pri Gorenjih Skopicah; 12- Velike Malence; 13- Veliki Korinj; 14- Šmihel pri Žužemberku; 15- Reva pri Dobrniču; 16- Mačkovec pri Dvoru; 17- Dolenje polje; 18- Groblje pri Novem mestu; 19- Beletov vrt; 20- Mačkovec; 21- Smolenja vas; 22- Otočec; 23- Brusnice; 24- Vinji vrh/Draga/Strmec; 25- Dolenje Gradišče; 26- Groblje pri Prekopi; 27- Brvi; 28- Stranska vas; 29- Verdun; 30- Mihovo; 31- Gradec nad Mihovim; 32- Gornja vas; 33- Hrast pri Jugorju – Grive; 34- Štrekljevec; 35- Borštek; 36- Rosalnice; 37- Otovec; 38- Otok; 39- Kučar; 40- Vrh pri Krasincu; 41- Griblje –Kohane; 42- Sadež; 43- Črnomelj; 44- Loka – Okljuk; 45- Tribuče; 46- Cerkvišče; 47- Griblje – Kamenica; 48- Šipek; 49- Gradenje; 50- Mekote; 51- Majišče; 52- Židovec; 53: Veliki Kolečaj

hillfort indicates the presence of a similar settlement or ecclesiastical centre on this site (Božič and Ciglenečki 1995: 266-267). The location of such centres is indicative of the continuing symbolic importance of these hillfort sites. Thus the construction of stone-built structures on prominent former hillforts, would also have symbolised the dominance of a new belief system over older belief systems, effectively incorporating important nodes in the ancestral landscape into the emerging landscape.

A final tantalising example of the re-use of Early Iron Age mortuary monuments can be detected in the 7th or 8th century AD. A single Early Slavic inhumation grave was inserted into a barrow in the Kozjane barrow group at the start of the ascent to the Vinji vrh hillfort (Mason 2001a: 24-25). The practice of re-using of earlier monuments in the Early Medieval period has been noted elsewhere in Europe. Slavic cemeteries are associated with Neolithic long barrows in Mecklenberg-Vorpommern in northern Germany (Holtorf 2000-2008), whilst in southern and eastern Britain Anglo Saxon cemeteries utilise Early Bronze Age barrows, Iron Age barrows and hillforts as well as Roman sites, as foci (Williams 1997: 1-32). Thus, we may have evidence of the use of Early Iron Age monuments to legitimate a new social order in the Early Medieval period in this part of Slovenia. However, given that Slavic burials may also be associated with other types of pre-existing monuments, such as the Roman settlement and cemetery at Dobova-Humek (Guštin 1982: 190-191; 1990: 17-20), the association with earlier sites may merely be an attempt to seek an association with something "old", whether Roman or Iron Age, to legitimate presence in the landscape.

The establishment of medieval churches on former hillfort sites could also symbolise the supplanting or Christianisation of earlier belief systems (Holtorf 1997: 80-88). This is particularly true of the establishment of an 11th century parish on the Vinji vrh hillfort (Peskar 1999: 32-49) (Fig. 2). The deliberate location of medieval churches on barrows, as at Čadraže (ANSl 1975: 220; Dular and Tecco Hvala 2007: 329) and Zemelj (Dular 1985: 85) (Fig. 3, Plate 2), is a potent symbol of domination over earlier belief systems through the monumental elements of the landscape.

Conclusion

In conclusion it may be seen that the Early Iron Age landscape of hillforts and barrow cemeteries in Central Slovenia represented in large part the monumentalisation of the pre-existing Late Bronze Age and possibly Middle Bronze Age habitation and mortuary sites, which often lay in dominant, highly visible positions in the landscape. Some of these earlier sites were in themselves seen as distant ancestral places, whilst others may have been directly linked to elite competition at the end of the Late Bronze Age and may thus have been directly related to the creation of the Early Iron Age socio-political milieu. The creation of monumental structures in the landscape employed or recycled recent material cultural to connect earlier sites to the living Early Iron Age society. Repeated use and incorporation invested the barrows and hillforts with symbolic meaning as icons of activity. Where such earlier places did not exist, earlier material was often incorporated into monuments to link them to the past. Such monumentalised places were often incorporated into the landscapes of the Late Iron Age,

Plate 2: *The church of Sv. Helena on the large Early Iron Age barrow at Zemelj (Photo: Ildikó Pintér).*

the Roman period and Early Medieval period. Thus, it is posited that the incorporation of earlier elements in a new whole was a means of legitimating a changing socio-political milieu through the recycling of, and as such, the reinterpretation of real and fictitious ancestral places.

Bibliography

ANSl *Arheološka najdišča Slovenije*, Ljubljana, 1975.

BOŽIČ, D.
 2008 *Late La Tène -Roman cemetery in Novo mesto: Ljubljanska cesta and Okrajno glavarstvo (Poznolatensko-rimsko grobišče v Novem mestu: Ljubljanska cesta in Okrajno glavarstvo)*. Katalogi in Monografiji 39, Ljubljana.

BOŽIČ, D. and CIGLENEČKI, S.
 1995 Zenonov tremis in poznoantična utrdba Gradec pri Veliki Strmici (Der Tremissis des Kaisers Zeno und die spätantike Befestigung Gradec bei Velika Strmica). *Arheološki vestnik 46*. Ljubljana, 247–277.

BREŠČAK, D.
 1990 Medvedjek, Trebnje. *Arheološka najdišča Dolenjske, Arheo 20*, Novo mesto, 43-44.

CIGLENEČKI, S.
 1987 *Höhenbefestigungen aus der Zeit vom 3. bis 6. Jh. im Ostalpenraum*. Dela 1. razprava SAZU 31, Ljubljana.

COLES, J. and HARDING, A.
 1979 *The Bronze Age in Europe*. London: Methuen

DULAR, A.
 1991 *Prazgodovinska grobišča v okolici Vinjega vrha nad Belo Cerkvijo (Die vorgeschichtlichen Nekropolen in der Umbegung von Vinji vrh oberhalb von Bela Cerkev)*. Katalogi in Monografiji 26, Ljubljana.

DULAR, J.
 1985 *Topografsko področje XI (Bela krajina)*. Arheološka topografija Slovenije, Ljubljana.
 1999 Ältere, mittlere und jüngere Bronzezeit in Slowenien - Forschungsstand und Probleme (Starejša, srednja in mlajša bronasta doba v Sloveniji - stanje raziskav in problemi). *Arheološki vestnik 50*, Ljubljana: 81-96.
 2001 Neolitska in eneolitska višinska naselja v osrednji Sloveniji (Neolitische und äneolitische Höhensiedlungen in Zentralslowenien). *Arheološki vestnik 52*. Ljubljana, 89–106.
 2008 Mihovo in severni obronki Gorjancev v prvem tisočletju pr. Kr.(Mihovo und die nördlichen Ausläufer der Gorjanci im ersten Jahrtausend v. Chr.) *Arheološki vestnik 59*, Ljubljana, 111–148.

DULAR, J. CIGLENEČKI, S. and DULAR, A.
 1995 *Kučar: Železnodobno naselje in zgodnjekrščanski stavbni kompleks na Kučarju pri Podzemlju (Eisenzeitliche Siedlung und frühchristlicher Gebäudekomplex auf dem Kučar bei Podzemelj)*. Opera Instituti Archaeologici Sloveniae 1, Ljubljana.

DULAR, J. and KRIŽ, B.
 2004 Železnodobno naselje na Cvingerju pri Dolenjskih Toplicah (Die eisenzeitliche Siedlung auf dem Cvinger bei Dolenjske Toplice). *Arheološki vestnik 55*, Ljubljana, 207-250.

DULAR., J., KRIŽ, B., SVOLJŠAK, D. and TECCO HVALA, S.
 1991 Utrjena prazgodovinska naselja v Mirenski in Temeniški dolini (Befestige prähistorische Siedlungen in der Mirenska dolina und der Temeniska dolina). *Arheološki vestnik 42*, Ljubljana, 65-205.

DULAR., J., KRIŽ, B., SVOLJŠAK, D. and TECCO HVALA, S.
 1995 Prazgodovinska višinska naselja v Suhi krajini (Vorgeschichtliche Höhensiedlungen in der Suha krajina). *Arheološki vestnik 46*, Ljubljana, 89-168.

DULAR, J. and TECCO HVALA, S.
 2007 *South-Eastern Slovenia in the Early Iron Age (Jugovzhodna Slovenija v starejši železni dobi)*. Opera Instituti archaeologici Sloveniae 12, Ljubljana.

GUŠTIN, M.
 1978 Gradišča železne dobe v Slovenije (Typologie der eisenzeilichen Ringwalle in Slowenien) *Arheološki vestnik 19*, Ljubljana, 100 -121.
 1982 Dobova. *Varstvo spomenikov 24*. Ljubljana, 190-191.
 1990 Dobova. *Arheološka najdišča Dolenjske. Arheo 20*, Novo mesto, 17-20.

HOLTORF, C.
 1997 Christian Landscapes of Pagan Monuments. A Radical Constructivist Perspective. In G. Nash (ed) *Semiotics of landscape: Archaeology of Mind*, BAR International Series 661, Oxford, 80-88
 2000-2008 *Monumental Past: The Life-histories of Megalithic Monuments in Mecklenburg-Vorpommern (Germany)*. Electronic monograph. University of Toronto: Centre for Instructional Technology Development. http://hdl.handle.net/1807/245.

HORVAT, M.
 2005 Loka. In: Djurić, B. (ed.). *The Earth beneath your feet: the archaeology on the motorways of Slovenia: guide to sites*.Ljubljana, 183–184.

KRIŽ, B.
 1989 Otočec. *Varstvo spomenikov 31*, Ljubljana, 213-214.

KRIŽ, B.
 1992 Arheološko območje Cvinger. *Varstvo spomenikov 34*, Ljubljana, 81 - 90,

KRIŽ, B.
 1997 *Kapiteljska njiva, Novo mesto*. Novo mesto: Dolenjski muzej

MASON, P.
 1996a *The Early Iron Age of Slovenia*. British Archaeological Report International Series 643, Oxford: Archaeopress.
 1996b Iron, Land and Power: The Social Landscape in the Southeastern Alps in the Late Bronze Age and the Early Iron Age. pp. 273-282. In: Jerem, E. and Lippert, A. *Internationales Symposium, Die Osthallstattkultur.*, Budapest: Archaeolingua.
 1998 Late Roman Črnomelj and Bela Krajina. *Arheološki vestnik 49*, Ljubljana, 285-313.
 2001a Družinska vas. *Varstvo spomenikov 38*. Ljubljana, 24-25.

2001b Griblje in problem nižinskih arheoloških kompleksov v Sloveniji. *Varstvo spomenikov 39*. Ljubljana, 7-27.

2005 Dolge njive near Bela Cerkev. In: Djurić, B.and Prešeren, D. (eds.), *The EARTH beneath your feet: the archaeology on the motorways of Slovenia: guide to sites*. Ljubljana: The European heritage days series, 123–125.

2006 Velike njive pri Veliki vasi. *Varstvo spomenikov 39-41 - Poročila*. Ljubljana, 230–231.

2008 Places for the Living, Places for the Dead and Places in Between: Hillforts and the Semiotics of the Iron Age Landscape in Central Slovenia. In: Nash, G. and Children, G. (eds.), *The Archaeology of Semiotics and the Social Order of Things*. British Archaeological Reports, International Series 1833, Oxford, 97–106.

2009 Place and Space in the Late Bronze Age and Early Iron of Central and Eastern Slovenia. In Nash, G.H. and Gheorghiu, D. (eds.) *The Archaeology of Territoriality and Space*. Budapest: Archaeolingua, 217-234.

2012 Something Old, Something New, Something borrowed ...: Romanisation and the Creation of identity in Early Roman central and south-eastern Slovenia.. In Lazar, I. and Županek, B. (eds.) *Emona: med Akvilejo in Panonijo - between Aquileia and Pannonia*. Koper: AnnalesMediterranei, 389-406.

MASON, P. and ANDRIČ, M.
2009 Neolithic/Eneolithic settlement patterns and Holocene environmental changes in Bela Krajina (south-eastern Slovenia). *Documenta Praehistorica 36*, Ljubljana, 327-325.

MASON, P. and BRITOVŠEK, T.
2007 Poročilo o arheološkem vrednotenju na območju Poslovno storitvene cone Mačkovec, unpublished report, ZVKDS OE Novo mesto.
2008 Poslovno storitvena cona Mačkovec: Poročilo o arheoloških izkopavanjij na območju PSCM 4, unpublished report, ZVKDS OE Novo mesto.

PESKAR, R.
1999 Cerkev sv. Jožefa na Vinjem vrhu pri Beli Cerkvi, *Varstvo spomenikov* 38, 32 – 49.

SHARPLES, N.
1991 *The English Heritage Book of Maiden Castle*, Batsford, London.

TECCO HVALA, S., DULAR, J. and KOCUVAN, E.
2004 *Železnodobne gomile na Magdalenski gori (Eisenzeitliche Grabhügel auf dem Magdalenska gora)*. Katalogi in monografiji 36, Ljubljana

TERŽAN, B.
1974 Halštatske gomile iz Brusnic na Dolenjskem (Die hallstattzeitlichen Grabhügel aus Brusnice bei Novo mesto). In Guštin, M. (ed.) *Varia Archaeologica*, Posavski muzej, Brežice, 33-66.
1999 An Outline of the Urnfield Culture Period in Slovenia. *Arheološki vestnik 50*: 97-143, Ljubljana

UDOVČ, K.
2009 *Mačkovec pri Novem mestu*. Zbirka Arheologija na avtocestah Slovenije 8, Ljubljana.

WILLIAMS, H.
1997 Ancient Landscapes and the Dead: the Reuse of Prehistoric and Roman Monuments as Early Anglo-Saxon Burial Sites. *Medieval Archaeology 41*, London, 1-32.

Recycling Pots, Places and Practices: The Roman Cemetery at Podlipoglav

Bernarda Županek and Irena Sivec

Abstract

In the paper we discuss a small Roman provincial cemetery, excavated in 1997 near the modern village Podlipoglav east of Ljubljana, Slovenia. This cemetery of a small hamlet in the backdrop of the colony of Emona came into use about a generation after the Roman conquest. Grave architecture and the structure of grave goods are a collage of different traditions, including Roman. The Podlipoglav cemetery and the sites in the wider area are strongly linked to a local prehistoric tradition, also through reuse of pre-Roman burial traditions and places in the Roman era. The recycling of pre-Roman traditions, as documented in the wider area of Podlipoglav, is in our opinion a marker of the importance that the predecessors had for the identity of the community inhabiting the area. We assume that the local community of the region was included in the Roman Empire through the formation of a new local, hybrid cultural identity, constructed through references to their ancestors and continuous tenure of the land.

Introduction: Recycling the Past

Only in recent years have archaeologists studied what has became known as "the past in the past" (Bradley and Williams 1998; Bradley 2002): different ways in which people would have inherited artefacts, settlements and whole landscapes from the past, ways in which ancient remains were invested with new meanings long after their original significance had been forgotten, how monuments were perhaps built to contrive the memories of future generations. Notions about how prehistoric life would always have been conducted according to an awareness of history that would have extended from the origins and use of the artefacts acquired in daily life, through the built environment that ancient people inherited, to the wider landscapes in which they lived, have only recently come into focus (Edmonds 1999; Bradley 2002).

A part of this wide area of study is the reuse of past in the past, or the recycling of the past (Mason, unpublished lecture). People in the past will always have been confronted by the surviving remains of antiquity that might be explained in many different ways (Bradley 2002), often by experience and expectations current at the time. Recycling of the past is a phenomenon recognized in a variety of archaeological contexts. It is employed through different means, combining objects, texts, oral traditions, iconographic representations and visible remains on the landscape.

Recycling of the past in the past has been interpreted in social, ideological and ritual terms. Used to construct meaning it is closely bound to the concept of social memory: the construction of a collective notion about the way things were in the past. Social memory is not about giving truthful and accurate testimony of past events, but about making meaningful statements about the past in a given present. This image or a set of images of the past is shared by a community, influencing their present and future actions. Often, the construction of collective memory and identity is highly politicised. Therefore, recycling of the past is rarely ideologically innocent, but often used to naturalize or legitimate authority, create and support a sense of individual and community identity or serve in resistance to imposed authority. Affirming a particular view of the past can also be a powerful method to resist change.

Pots, Places, and Burial Practices in the Podlipoglav Area

The Roman cemetery of Podlipoglav is located near the modern village of Podlipoglav, approx.10 km east from Ljubljana, Slovenia. Geographically, Podlipoglav is a part of a distinct region named the Posavsko hribovje. This hilly region, covered with woods and abundant in surface water, sharply ends to north with the Ljubljana Moors, on west with the Dolenjska lowlands, where the main Roman road between Italy and the Balkans ran (Fig. 1). To northeast, the Posavsko hribovje is cut by the river Sava, and towards the northwest a distinctive border is formed by the low and fertile Sava plain. Thus defined, the Podlipoglav area is connected with Ljubljana area and with the Zasavje and Dolenjska regions by a network of roads and paths. The area is an archaeologically distinctive region both in prehistory and in the Roman period (Sivec and Županek 2013). Unfortunately, available data derives mainly from old excavations and stray finds.

The wider area of Podlipoglav (Fig. 1) is covered by a dense network of prehistoric sites. In nearby village Pance, a prehistoric hillfort and Early Iron Age

Working with the Past: Towards an Archaeology of Recycling

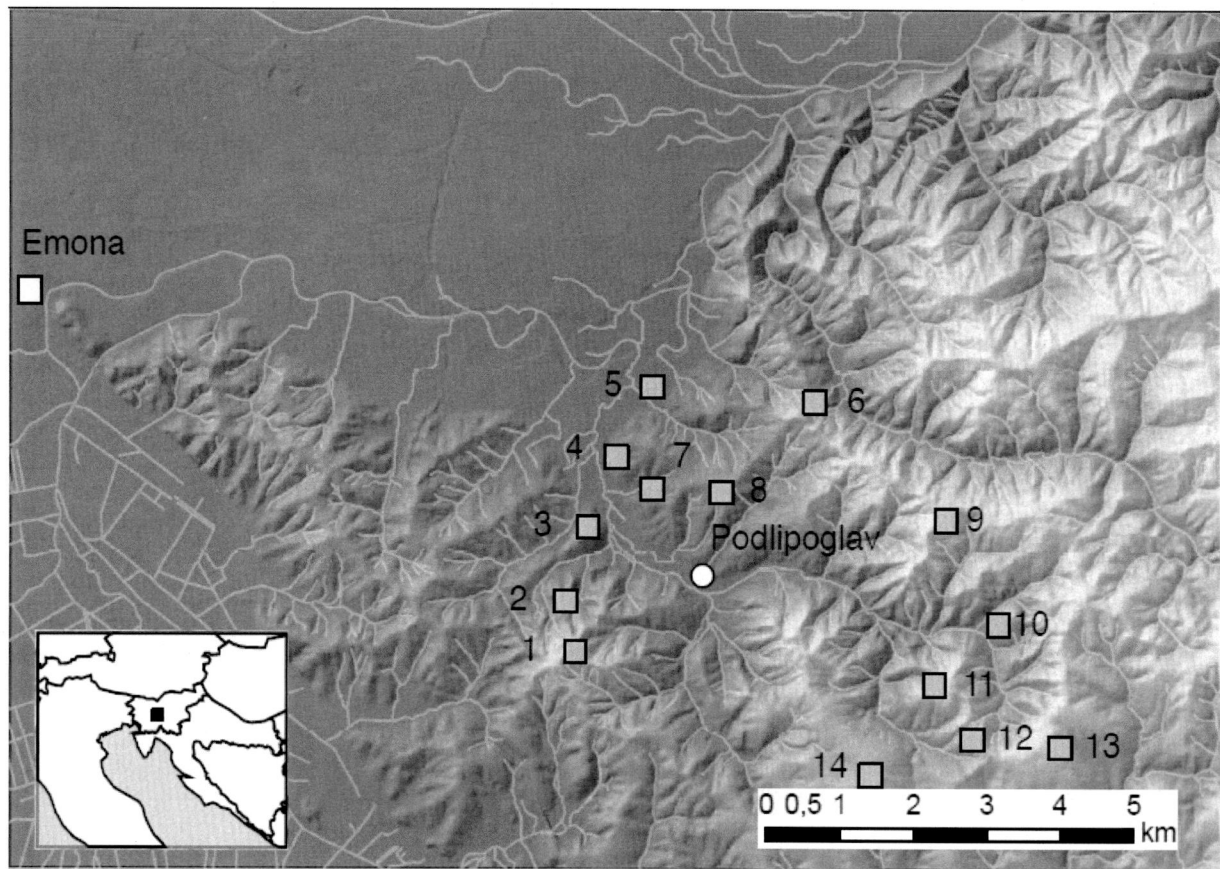

Fig. 1: *The Podlipoglav area with sites mentioned in the text: 1 Molnik, 2 Podmolnik, 3 Marenček, 4 Sostro/sv. Lenart, 5 Zavoglje, 6 Besnica/Tomaž, 7 Češnjica, 8 Gradišče/Zagradišče, 9 Javor, 10 Ravno brdo, 11 Veliki Lipoglav/Roje, 12 Mali Lipoglav/Mrdiž, 13 Pance, 14. Magdalenska gora.*

cemetery were found (Gabrovec 1975c; Dular 2003: 154). The villages of Mali and Veliki Lipoglav are also archaeological sites, the former with a rampart (Gabrovec 1975b) and the latter as a site where iron slag and barrows were found (Gabrovec 1975d). Magdalenska gora is an Early Iron Age hillfort with a barrow cemetery, and Molnik as well (Puš 1991). In the Late Iron Age several graves were cut into the existing Early Iron Age barrows at Magdalenska gora (Gabrovec 1975a: 201). There is also a Late Iron Age grave in an Early Iron Age barrow at Molnik (Puš 1991: 47). The hillforts at Marenček (Puš 1990), Češnjica at Sostro (Puš 1991: 52) and Gradišče at Zagradišče are tentatively dated to the Early Iron Age.

The area is well known for its iron, lead and zinc ore deposits (Drovenik et al. 1980: 20), which were exploited in prehistory (Tecco Hvala et al. 2004: 109s). Roman trade with finished products came early to this area, probably already in the Republican period, and was in hands of *liberti* from Aquileia and the Caesernii and Barbii families (Petru 1964-65: 78; cf. Šašel 1992).

The Roman conquest of the Podlipoglav area took place in the late 1st century BC and in the first decade of the 1st century AD. In the first decade of 1st century the *Colonia Iulia Emona* was built, being completed in AD 14/15. Cemeteries predominate in the archaeological record of the Roman period in the Podlipoglav area: Roman graves (2nd-4th centuries) were cut into an Early Iron Age cemetery in the nearby village Javor (Guštin and Knific 1973: 840); Roman burials were interred in Early Iron Age barrows at Magdalenska gora (Vuga 1988: 13), and a late Roman cemetery (4th century or later) was found at Ravno brdo (Stare 1952). In addition to these sites, Malo Trebeljevo is mentioned as a site with Roman walls and graves (Pečnik 1904: 130).

To summarise, the cemetery at Podlipoglav is located in a distinctive hilly area with a dense network of prehistoric sites and relatively few (in comparison with surrounding areas) Roman sites, some of them clearly connected with Early Iron Age sites.

Hybrid Identities: Mix and Match

The cemetery at the village of Podlipoglav was partly excavated in 1997 (Fig. 2). 33 graves were uncovered. Cremation is the exclusive burial rite, as in the Late Iron Age burials in the area. Three types of mortuary structures were found: a simple grave pit (in some instances covered with a stone slab, as was customary

Fig. 2: *The Podlipoglav cemetery under excavation.*

for the Early Iron Age graves in the barrows at nearby Molnik); rectangular grave pits, lined either with stone (as customary in the Celtic Latobici graves in the Dolenjska region) or with *tegulae* and burials in stone-walled grave pits.

The cemetery probably belonged to a small hamlet in the hinterland of the Roman colony of Emona and lay approximately 10 km from the city. At this time the area of Podlipoglav was located away from the Roman main road network and good arable soil, in an area where woods and marshy floodplains predominated, probably outside the *ager divisus*. The cemetery came into use about 50 years after the Roman conquest of the wider area. On the basis of the pottery assemblage, it has been traditionally interpreted as Roman, with strong prehistoric traits (Gaspari 1998).

The pottery assemblage consists mainly of vessels, some imported from northern Italy, some recognized as production from Emona and some made at home or locally. Some features, such as a distinct combination of drinking vessels (Fig. 5) is similar to the Celtic Latobici graves in the adjacent Dolenjska region. Several of the locally made jars are made in the local Iron Age tradition, reflected in the shapes, fabrics and firing techniques used.

On the basis of the pottery assemblages the use of the cemetery is dated from circa AD 50 to circa AD 400; there are no graves that could be firmly dated before AD 50. The majority of the graves date to the second half of the 1st century and the first half of the 2nd century. There is a radical decline in the number of graves after AD 200 (Fig. 3). Two or three rows of graves could be identified, composed of later graves (Fig. 4). Several graves with mortared stone walls are located on the southern edge of the cemetery, overlooking the floodplain and possibly a minor road.

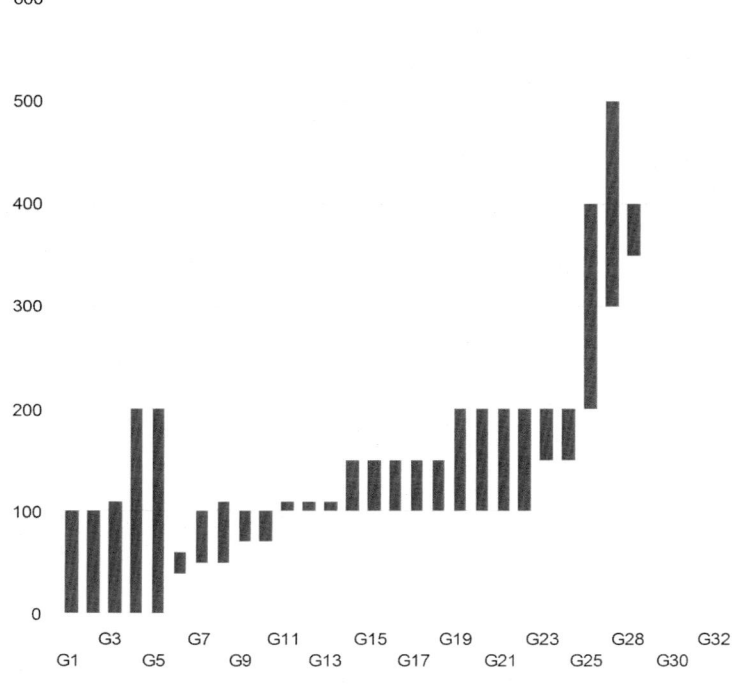

Fig. 3: *Dated graves.*

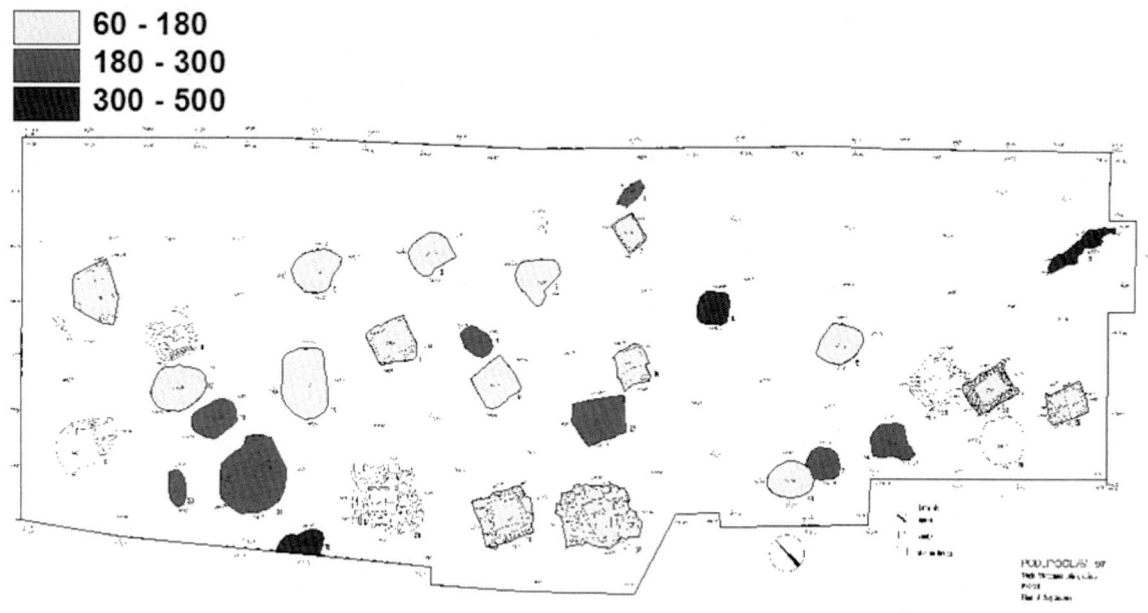

Fig. 4: *Plan of the excavated cemetery; a suggested chronology of the burials.*

We argue that the grave architecture and structure of the grave goods at the Podlipoglav cemetery are an eclectic blend of different traditions, usually labelled "indigenous", "Celtic", "Latobican", "local", "Emona", "imported", "foreign", "Roman" (Figs. 5-7). We suggest that the burial rite, mortuary structures and grave goods at Podlipoglav were structured through the process of bricolage.

Bricolage is a concept describing a process in which new cultural items are obtained by means of attributing new functions to previously existing ones (Terrenato 1998: 23s). Not only objects, but practices, rituals, ways of life can likewise be changed through bricolage: partly of traditional, recycled, and partly of new ways of doing things, working in new contexts. The outcome of bricolage resembles a collage: a complex patchwork made of elements of various age and provenance: some of them are new, but many others are old objects, refunctionalised in new forms and made to serve new purposes within a new context: and this, we think, in a plausible explanation of our situation at Podlipoglav and its wider area. Besides, bricolage, concept formed by Claude Levi-Strauss (1962) and frequently used in anthropology for exploring indigenous societies in colonial contexts, is, we think, particularly useful in Roman archaeology, as it enables us to surmount simple dichotomies, such as indigenous/newcomer, old/new, Roman/non-Roman. In short, bricolage prompts us to look at cultural and social change in the Roman period from a different viewpoint.

Using bricolage as a starting point we are steered towards another useful concept that explains the accommodation of new objects and practices into existing culture - that of cultural hybridity (Webster 2001):

Fig. 5: *The grave goods in Grave 2: combination of pot and cup in Latobican tradition, imported oil lamp, plate with red slip (imitation of sigillata ware), two vessel fragments. Photo Matevž Paternoster, MGML archive.*

Fig. 6: *Grave 2 during excavation. Photo Andrej Gaspari, MGML archive.*

Fig. 7: *Grave goods in grave 31. Photo Matevž Paternoster, MGML archive.*

a multicultural adjustment through which a new, mixed culture is formed. This concept, as the previous one, sees culture in a constant process of translation, beyond excluding, fixed binary identities. Therefore it is not possible to seriously speak about Early Iron Age traits, Roman traits, and prehistoric traits, nor to dissect the assemblage in these terms, or claim that some traits are more authentic than others. When at work, cultural hybridity takes hold of all the parts. It is a process of blending, through which a new, mixed, hybrid culture is formed. The blended, mixed, hybrid cultural sets that formed under Roman rule, including the one at Podlipoglav, are traditionally termed "Roman provincial culture". This is a broad umbrella term covering a wide range of different sets, different responses to different local situations.

Both bricolage and cultural hybridity seem appropriate models for understanding our case study. However, we are interested in the meaning conveyed by the chosen set of cultural references, the local bricolage – what were the people communicating in this fashion? What is the meaning of sets of objects and practices, traditionally regarded as belonging to at least two different cultures, Roman and indigenous, yet combined and bound firmly together through the burial?

As an important indicator towards an answer we consider the choice of cemetery location. When new graves were dug during the Early Roman period, they did not fill empty or untouched spaces, but were fitted into a landscape of ancient sites, which were still meaningful. The cemetery at Podlipoglav is located on a visible terrace above the floodplain, near the route towards the Zagradišče and Česnjica hillforts (Puš 1991: 52), where presumably the part of the community had lived until a generation ago. The main road ran on the other side of the valley, over the marshy floodplain. Marshes, liminal spaces, outside everyday use, places of special practices and stories, might also be a – perhaps metaphorical – border for the community. The cemetery area was probably also chosen with a desire for good visibility from the road. The graves were obviously clearly visible on the surface and perhaps marked. A significant number of graves are lined with mortared stone walls (Fig.6), indicating the time and resources that were invested in grave construction. The cemetery was located on a good arable soil, an important message in a region where arable soil is scarce and woods, hilly pastures and marshy floodplains predominate.

Additionally, Podlipoglav is a part of an archaeologically distinct region with two known instances of reuse of Early Iron Age cemeteries. The ancestors and the past were obviously important for the communities that lived here. How and in what way?

Reuse of the Past: "Recycling the Ancestors" at Podlipoglav

The cemetery at Podlipoglav is located in a distinctive hilly area with a dense network of prehistoric sites and relatively few (in comparison with other areas) Roman sites, some of them clearly connected with Early Iron Age sites. There are at least two instances of reuse of pre-Roman burial places in the Roman period in the area: Javor (Guštin and Knific 1973), with Roman graves in the Early Iron Age cemetery, and Magdalenska gora, a large Early Iron Age hillfort on the border of Posavsko hribovje, above the important communication route, with Roman graves cut into the Early Iron Age barrows. To understand this, we use the idea of "recycling the ancestors", as developed by Phil Mason in his paper at the EAA meeting in 2009. In his words, we can understand such reuse as: "recycling of certain ancestral places and practices in the context of the creation of new local identities, based on identification with existing loci in the landscape in the wake of the reorganisation or collapse of traditional systems of power".

The Podlipoglav area clearly reflects a preoccupation with the past through the interment of their deceased in earlier cemeteries in Javor and Magdalenska gora. To say that Early Iron Age burial places were reused is one thing; to say that cemeteries were reused after 20-30 generations is quite another. It suggests that at least some traditions were remarkably persistent and we should ask why that might be. Were ancient cemeteries visible monuments? Was the intended message that of the ancestral connections of the community and their tenure of the land? This might be one of the answers, as the location for the new cemetery was clearly carefully selected: it was visible from the route ways below, and close to an old communication to the now abandoned hillforts, associated with the ancestral past. To gain good visibility for the new cemetery well-drained land in this rather marshy area was removed from cultivation, which also suggests that the community associated with it wished to emphasize the importance and extent of their rights over the land in the area. The location of the cemetery in a liminal place at the beginning of the marshland might also indicate a community border.

Were the inhabitants referring to the duration of their community and ancestors in an area that was occupied by the Romans a generation before and the existing relations of power redefined? Are the secondary burials in ancient mounds an attempt to gain prestige and status by manipulating genealogies or inventing ancient traditions? By burying their dead in ancient tombs, could they have been constructing myths of origin and relations with distant past?

Becoming Roman at and around Podlipoglav: A New Story of Old Times?

To conclude, the cemetery at Podlipoglav is located in a distinctive hilly area with a dense network of prehistoric sites and relatively few Roman sites. Some cases of reuse of Early Iron Age cemeteries are known. The excavated cemetery at Podlipoglav was the cemetery of a small hamlet. The first generation after the Roman conquest buried their dead elsewhere (probably around the Gradišče hillfort directly above Podlipoglav). Grave constructions and grave goods structure show an eclectic blend of different traditions. Among the objects and practices used for this bricolage, several can be interpreted as items of an unbroken tradition or perhaps intentionally recycled from the early and Late Iron Age.

Podlipoglav is a cemetery of a local population that used a wide range of objects and burial architecture, including Roman. We would like to emphasize the obvious importance of predecessors for identity of the community of the wider area of Podlipoglav, reflected both in use of old cemeteries, construction of graves, pottery assemblages objects, places, meanings and memories that were intertwined in a community-specific sense of place. These practices seem to show commitment to traditional values, probably important for the community's social identities at the time. Indeed the area of Podlipoglav, with its reuse of Early Iron Age burial places, indicates that burial rites and mores might be frames upon which social memories and values could be inscribed. Were these a myth of distant origins, a *sui generis* myth, crucial for the redefinition and reproduction of local authority in the times of change? Associating with a powerful distant past by reusing ancient burial grounds may have served to legitimise political strategies in the present, symbolising links to ancestors and trying to emphasize its claims to the land, deriving from these ancestors, as suggested for some similar cases in Dolenjska by Phil Mason (Mason, unpublished lecture).

The historical context for the reuse of the past in the wider Podlipoglav area might well be the restructuring of old/emergence of new political structures at in the decades after the Roman conquest. Emphasis on the past would have been particularly appropriate during a period in which new regions were settled and new relationships were formed between people and the land. Whether this referencing of the past, the ancestors, was continuous or newly invented is less important than the likelihood that a reference to ancestral claims was intended.

We presume that the community in the area under discussion was integrated into the Roman Empire through the dynamic formation of a new, local, hybrid cultural identity, characterised by referring to ancestors and ancient tenure of this land, perhaps exploiting an origin myth for political advantages. We suggest that the recycling of the past in the wider area of Podlipoglav could be understood in the context of inventing a specific cultural tradition (cf. Hobsbawm and Ranger 1983) during the social and political transformation at the beginnings of Roman dominance in this area.

In any case, we believe that the reuse of the prehistoric cemeteries and the recycling of the past in the Roman period have far-reaching implications for the ways in which we think about Roman rural settlement and the groups who lived and were buried there.

Acknowledgements

We would like to thank Dragoş Gheorghiu and Phil Mason for inviting us to the 2009 EAA session *Working with the past: strategies for crisis or intentional incorporation? Towards an archaeology of recycling* and for inviting us to write this paper.

The responsibility for the content of this paper lies with us alone.

Bibliography

BRADLEY, R.
2002 *The past in prehistoric societies.* Routledge: London, New York.

BRADLEY, R. WILLIAMS, H. (ed.)
1998 The past in the past: the reuse of ancient monuments. *World Archaeology* 30/1.

DROVENIK, M., PLENIČAR, M., DROVENIK, F.
1980 Nastanek rudišče v SR Sloveniji. *Geologija* 23/1: 1-157, Ljubljana.

DULAR, J.
2003 *Halštatske nekropole Dolenjske.* Opera Instituti archaeologici Sloveniae, 6. Ljubljana: Založba ZRC.

EDMONDS, M.
1999 *Ancestral geographies of the Neolithic. Landscapes, monuments and memory.* Routledge: London, New York.

GABROVEC, S.
1975a Magdalenska gora. In: *Arheološka najdišča Slovenije.* Ljubljana, Državna založba Slovenije, 200-201.
1975b Mali Lipoglav. In: *Arheološka najdišča Slovenije.* Ljubljana, Državna založba Slovenije, 201.
1975c Pance. In: *Arheološka najdišča Slovenije.* Ljubljana, Državna založba Slovenije, 201-202.
1975d Veliki Lipoglav. In: *Arheološka najdišča Slovenije.* Ljubljana, Državna založba Slovenije, 202.

GASPARI, A.
1998 Podlipoglav. *Varstvo spomenikov*, 38: 87-88. Ljubljana.

GUŠTIN, M. and KNIFIC, T.
1973 Halštatske in antične najdbe iz Javora. *Arheološki vestnik*, 24: 831-841, Ljubljana.

HOBSBAWM, E. and RANGER T. (eds.)
1983 *The Invention of Tradition.* Cambridge: Cambridge University Press.

KNEZ, T.
1992 *Novo mesto II. Keltsko-rimsko grobišče Beletov vrt.* Carniola Archeologica 2. Novo mesto, Dolenjski muzej.

LEVI-STRAUSS, C.
1962 *La pensée sauvage.* Paris: Plon.

MASON, P.
unpublished lecture, Recycling the ancestors: Roman cemeteries and the reuse of prehistoric mortuary monuments in central and south-east Slovenia. Lecture given at 14th Annual Conference of the European Association of Archaeologists, Malta, 19.9.2008.

PEČNIK, J.
1904 Prazgodovinska najdišča na Kranjskem. *Izvestja muzejskega društva za Kranjsko*, 14: 130, Ljubljana.

PETRU, P.
1964-65 Nekateri problemi provincialno rimske arheologije v Sloveniji. *Arheološki vestnik*, 15-16: 65-97, Ljubljana.

PUŠ, I.
1990 Mareček – Višinska postojanka in refugij. *Arheološki vestnik*, 41: 365-374, Ljubljana.
1991 *Molnik: sedež prazgodovinskih knezov.* Ljubljana: Mestni muzej.

SIVEC, I. and ŽUPANEK, B.
2013 Stiks na podeželju: grobišče pri Podlipoglavu blizu Ljubljane. *Studia Universitatis Hereditati* (1-2): 11-24.

STARE, V.
1952 Ravno brdo. *Arheološki vestnik* 3/1: 137-144. Ljubljana.

1992 Caesernii. In: Bratož, R., Šašel Kos M. (eds.) *Opera selecta. Situla, Razprave Narodnega muzeja v Ljubljani*. Ljubljana: Narodni muzej Slovenije, pp. 54-74.

TECCO HVALA, S., DULAR, J., KOCUVAN, E.
2004 *Železnodobne gomile na Magdalenski gori*. Katalogi in monografije, 36. Ljubljana, Narodni muzej Slovenije.

TERRENATO, N.
1998 The Romanization of Italy: global acculturation or cultural *bricolage*? In: Forcey, C., Hawthorne, J., Witcher, R. (eds.) TRAC *97, Procceedings of the Seventh Annual Theoretical Roman Archaeology Conference*, Oxford: Oxbow books, pp. 20-27.

VUGA, D.
1988 *Magdalenska gora pri Šmarjem-Sapu v železni dobi*. Kulturni in naravni spomeniki Slovenije, Zbirka vodnikov 157. Ljubljana, Založba Obzorja.

WEBSTER, J.
2001 Creolizing the Roman Provinces. *American Journal of Archaeology*, 105: 209-225. Boston, Archaeological Institute of America.

Secondary Use of Storage Vessels and Household Pottery During the Late Middle Ages: Pottery in Vaults as a Case Study

Marta Caroscio

Abstract

The use of amphorae and other vessels for buildings during Late Antiquity and the Middle Ages is very well known but different from the Roman period. When talking about this building technique Vitruvius (V, 5) refers to vases used for acoustic purpose. Even though during the 5th and 6th centuries some technical devices changed, and despite the examples in the western Mediterranean are scarce between the 6th and 10th centuries, vessels were still placed in vaults during the Late Middle Ages. Recent studies have drawn attention on the variety of technical devices used (Poisson 2005, Berti 2007), pointing out that during the late Middle Ages the vessels employed as vault fill were mainly "recycled" (kiln, trade or domestic waste). Were local products only used? What was the role of storage vessels? How these assemblages have been interpreted?

This paper aims to present part of the results of a broader research on the production and use of storage vessels in the western Mediterranean between the 14th and the 17th centuries (Caroscio 2009). Diverse examples of secondary use of pottery for building vaults will be discussed, presenting, among others, the assemblage of the "Convento de Santo Domingo" (Valencia) as a case study.

This paper aims to present part of the results of a broader ongoing research on the production and use of storage vessels in the western Mediterranean between the 14th and the 17th centuries (Caroscio in press). Diverse examples of secondary use of pottery for vaults filling will be discussed, presenting, among others, the assemblage of the "Convento de Santo Domingo" ("Monastery of Saint Domingo", Valencia) as a case study. The use of amphorae and other vessels in architecture was extremely common during the late antiquity; but during the Middle Ages recycled materials only were used (Fichera 2007: 154). As pointed out by Poisson (2005: 58), when referring to vessels employed in vault fill, Vitruvius (V, 5)[1] mentioned especially the use of vases for acoustic purposes.[2]

Even though most of the vessels employed in architecture during the Roman period were made on purpose, cases of "recycled" amphorae used as vaults filling are known since the 5th century. We can mention, for example, the church of San Vitale in Ravenna and the church of San Simpliciano in Milan (Vannini 2001: 199). In both cases we are talking about the secondary use of amphorae previously employed for trading goods and then recycled as vaults filling. Thus, it can be said that the technique changed during the 5th and the 6th centuries, as filling materials were not produced on purpose any longer, but become more and more common to use kiln waste or waste of domestic households (Berti 2007: 154). According to the evidence known so far, there is a kind of gap in the use of this technique, but it cannot be assumed for sure that it was not used at all between the 7th and the 9th century. The inclusion of the well-known text *Magister Commacinus* in the "Edict of Rotari" (22nd November 643) underlies the idea that the technique was known in the 7th century. More than half a century later, the "Edict of Liutprando", still includes an appendix with a *Memoratium de mercedibus commacinorum*. When describing how to build a kiln, the first text clearly mentions the use of vessels (*caccabos*) for filling the vault, so to make it lighter (Monneret de Villard 1920: 10).[3]

---

[1] *Ita ex his indagationibus mathematicis rationibus fiant vasa aerea pro ratione magnitudinis theatri eaque ita fabricentur ut cum tangantur, sonitum facere possint inter se diatessaron diapente et ex ordine ad disdiapason* (V, 5.1). *Ita hac ratiocinatione vox a scaena uti ab centro profusa se circumagens tactuque feriens singulorum vasorum cava excitaverit auctam claritatem et concentu convenientem sibi consonantiam* (V, 5.3). *In medio nihil est conlocandum, ideo quod sonituum nulla alia qualitas in chromatico genere symphoniae consonantiam potest habere. In summa vero divisione et regione cellarum in cornibus primis ad diatonon hyperbolaeon fabricata vasa sonitu ponantur* [...], (V, 5.5).

[2] The use of acoustic vessels is not going to be discussed in this paper. For the Middle Ages see: Arns and Crawford 1995. A research on the use of acoustic vases in late antiquity has recently been presented at the AIECM2 Conference in Venice (Cuteri & Di Fede 2009).

[3] *Si quis vero furnum in pisile cum caccabos fecerit, et postes tres aut quatuor habuerit, et cum pineam suam levaverit caccabos ducenti quinquaginta, ita ut pinea ipsa habeat caccabos viginti quinque exinde tollent tremisse unum. Et si quingenti caccabos habuerint, habeant duos tremisse unum. Et si mille fuerint caccabos, tollent exinde mercedes tremmisse quattuor* (Memoratorium de Mercedibus Commacinorum, IX de furnum, in Monneret de Villard 1920: 16; after Blume 1868).

Turning to archaeological evidence, one of the earliest examples, dating to the 10th century, seams to be the cathedral of Saint Donatian in Bruges. Nevertheless, the technique was wide spread in the Mediterranean area rather than in northern Europe. Recent studies have actually drawn attention on the variety of technical devices used (Poisson 2005, Berti 2007), pointing out that during the Late Middle Ages vessels employed for this purpose were mainly "recycled" (kiln, trade or domestic waste). Were local products only used? What was the role of storage vessels? How these assemblages have been interpreted? Before answering these questions, it should be pointed out that different techniques were used for filling vaults with vessels during the Late Middle Ages. Moreover, the vessels used were not made on purpose; nevertheless, they were quite "standardized". Even though big storage vessels were normally employed for this purpose in the western Mediterranean, cases of secondary use of tableware and kitchenware are known all around Tuscany during the 14th century. Among those it is worth mentioning San Domenico in Prato (Vannini 2001), Sant'Antimo sopra i canali in Piombino (Berti & Bianchi 2007) and Santa Maria del Carmine in Siena (Francovich & Valenti 2002).

If we compare the different techniques and the diverse materials used for vaults filling in the Middle Ages and in the Roman world, it shows up that during the Roman period objects from long distance trade are more common, while during the Late Middle Ages most of the pottery used was of local production. Moreover, the way itself of filling vaults was different, as more layers were used in the Middle Ages. Each layer was filled in with certain kinds of vessels, so that they could fit the shape of the vault. Santa Maria della Scala in Siena is a clear example of the use of this technical device (Francovich & Valenti 2002).

Concerning recycling, both domestic vessels that had lost their primary function and kiln waste were normally used for this purpose. Trying to answer the question about the circulation of pottery, it can be said that domestic vessels usually came from the local area. This means that they could be either local productions or imported objects consumed locally. Turning to analyse kilns wastes, it is worth noting that they were bought from local workshops, but in some cases they were traded within the regional area. So, what was the role of storage vessels? Could it be somehow compared to that of Roman amphorae? Big vessels used for storing food or goods could be produced locally, but quite often they were imported from the western Mediterranean, especially from the eastern coast of Spain. Nevertheless, archive sources give evidence of waste bought from local kilns and archaeological excavation have confirmed this data. Examples are known not only in Tuscany, but also all around the western Mediterranean.

One of the most interesting cases of "recycling" kiln wastes for filling vaults is recorded during the construction of the cathedral in Barcelona. Records of all the expenses on the building site are kept from 1298 up to 1445. The first document referring to the building of the vaults of the cathedral in Barcelona dates back to the 1379.[4] A few years later (1381) a similar document states clearly how much did the vessels coast, but in spite of this fact, there is not clear reference to secondary use of kiln waste. Nevertheless, the amount paid seems to refer to cheaper objects than those normally sold, and some of them were actually donated by the potters, possibly outlining that those vessels might be somehow flawed.[5] The first clear reference to "broken" vessels (*frentum*) in the account books of the cathedral dates to 1418; at that time some large storage vessels were bought and the document clearly explains that *frentum* meant "broken" (*trencades*) or flawed vessels.[6] Reference to secondary use of storage vessels is not unique and some more jars were bought later on during the same year as well as in the following ones.[7]

A similar use of big storage jars is recorded more than one century later on the other shore of the western Mediterranean. In 1563 Virgilio Carnesecchi, when reporting to the Great Duke Cosimo on the expenses for building the Medici Palace in Seravezza, listed big jars (*coppi*) among the materials bought.[8] In the same regional area and during the same period, the vaults of the monastery of Sant'Agostino in Pietrasanta (Berti 1990) were filled in with Spanish jars and tin-glazed pottery from the Valencian area (Berti -Tongiorgi 1974; Francovich & Gelichi 1984).

If we take into account the latest archaeological evidence of vaults filling recorded in northern-central Italy, at least two cases should be mentioned: one is the church of "Sant' Antimo sopra i canali" in Piombino

---

[4] Referring to the vault of the nave: paguí per gerres que en Pere Alegre compra als ollés obs de la volta (Carreras y Candi 1913: 132).
[5] Item paguí per gerres que en Pere Alegre compra als ollers, obs de la volta, 5 sous i 6 diners. Item per port de dites gerres i d'altres que ens donaren, 4 sous i 6 diners [9th June 1380]; (Bassegoda Nonell 1983: 64; with Pera Alegra corrected into Pere Alegre).
[6] Compri den verger IIII somades de frentum, ço és, canters, olles, terrassos e alters fraschas de terra trenchats, costàren a raó de XII diners per somada [...] dating 20th June 1418 (Carreras y Candi 1914: 311).
[7] On the 22nd October 1418: lo dia mateix paguí en Johan Verger, gerrer, per la volta major, a raó de 4 diners per peça e comprim 26, qui muntaren a la dita raó sous, 8 diners. Item comprí de dit Sebastià, gerrer, per la dita volta 4 somades de frentum, qui a raó de 1 sou, 6 diners per somada, munten sis sous (Bassegoda 1983: 65); the 1st December 1431: "y a més, compra de frentum o gerres y canters trencats, per posar dintre les voltes" (Carreras y Candi 1914a: 303); and again the 21st May 1435: XVIII geras trencades per la volta de la claustra (Ivi: 304).
[8] Item p 120 coppi p mettere alle volte_____scudi 20 (Berti & Tongiorgi 1974, after Buselli 1965: 196).

(Berti & Bianchi 2007) and the other one is "Santa Maria del Carmine" in Siena (Francovich & Valenti 2002). In Piombino (southern Tuscany) kiln waste traded from Pisa was used as vault fill together with domestic waste. It has been suggested that these vessels were shipped in there for building purposes as no local production has been recorded so far (Berti & Bianchi 2007). If this was the case, as it seams to be likely, it should be beard in mind that pottery was not shipped alone, as it was only a minor part (and usually one of the less valuable objects) among the diverse goods that were shipped (Molinari 2003). Concerning the interpretation of this assemblage it might be possible that different kind of wastes (domestic and kiln waste) were collected in Pisa, possibly in different areas of the town, rather than assuming that the domestic waste was collected in the environs of Sant'Antimo. In several excavations carried out in Pisa during the 1960s and the 1970s, assemblages of kiln waste dating to the 15th-16th centuries were recovered below the floor of a house (Berti 2007: 374; Berti 2005: 5-8). Unfortunately, no written evidence of the trade of this objects from Pisa to Piombino is known so far (Berti & Bianchi 2007). An earlier example is Santa Maria del Carmine in Siena, dating to the 13th-14th centuries. This assemblage confirms the combined use of domestic waste and waste from local kilns (Francovich & Valenti 2002) as still in use a couple of centuries later like in the context analysed above.

The case study analysed in Valencia is the assemblage of storage jars (*tinajas*) recovered in the vault fill of the "Iglesia de Santo Domingo" and dating to the late 14th-15th centuries (Dies Cusi & Gonzales Villaezscusa). Some of these vessels were kiln waste from local workshops, while another part were domestic waste of imported jars, possibly containing oil, traded from southern Spain, especially from Seville (Fig. 1-3). It is worth noting that while kiln wastes consisted of local productions, the secondary use of jars refers both to vessels manufactured locally and to imported ones (Amigues *et alii* 1996). Santo Doming is not a unique case: the trade of kiln waste and second hand objects, re-used after their primary function was over, is quite common in the Catalan area during the 14th century (Gonzalez Gonzalo 1987: 479).

Summing up, it can be stated that even though the technique of filling vaults with empty vessels derives from Roman building techniques, it underwent changes during Late Antiquity. Despite archaeological evidence are lacking between the 7th and the 9th centuries, written sources like the *Magister Commacinus* seams to underlie a transmission of the technique. The fact that Vitruvius refers to acoustic vases should not be regarded as conflictive evidence, because the technique possibly spread after the treaty was written (Poisson 2005: 58), especially from the 5th century onwards. From then onwards, large storage vessels

Fig. 1: *"Iglesia de Santo Domingo" (Valencia, Spain), secondary use of kiln waste as vaults filling.*

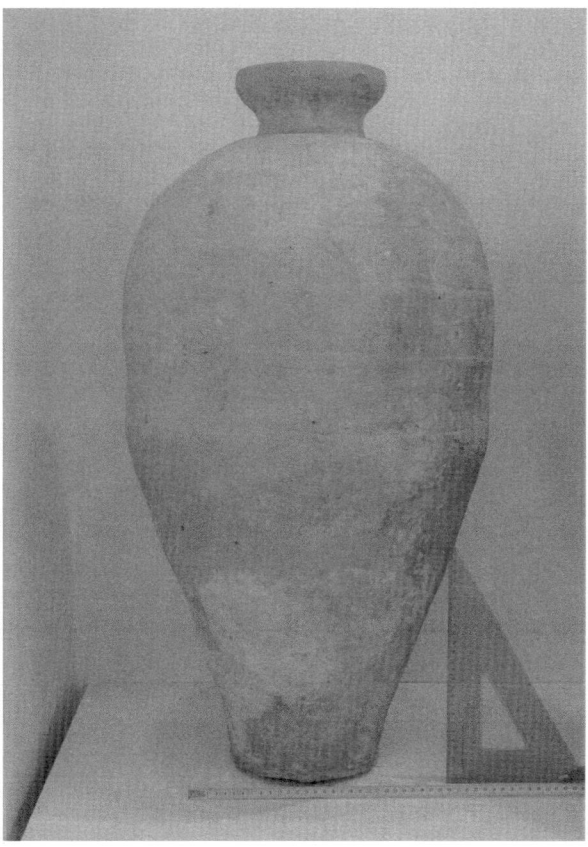

Fig. 2: *"Iglesia de Santo Domingo" (Valencia, Spain), secondary use of a jar locally produced and used (domestic waste) as vault fill.*

and amphorae were clearly used to make the structure lighter. While during Late Antiquity the material employed in buildings consist mainly of standardized

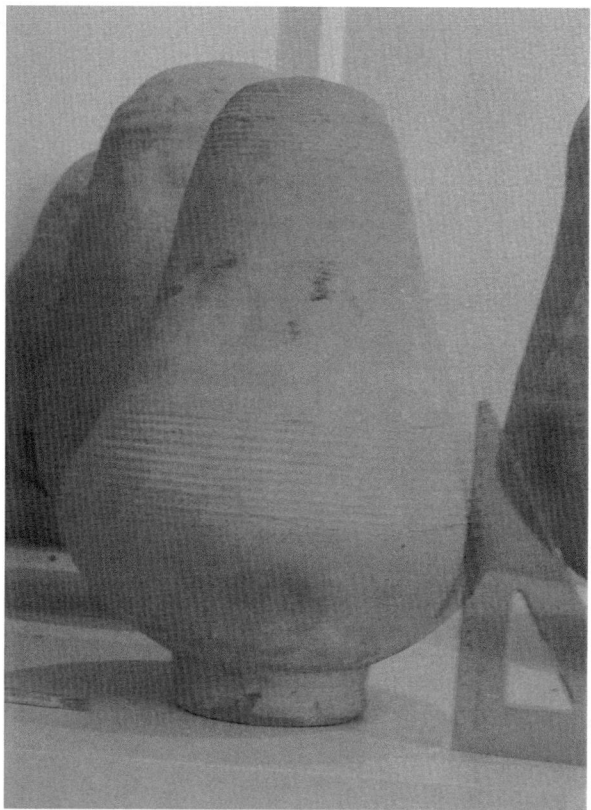

Fig. 3: *"Iglesia de Santo Domingo" (Valencia, Spain), secondary use of an imported jar from Seville as vault fill.*

clay pipes made on purpose, the vessels used during the Middle Ages were "recycled" and could be either kiln or domestic waste. Nevertheless, quite a few examples of secondary use of amphorae are known during the 5th and the 6th centuries (see San Vitale in Ravenna and San Simpliciano in Milan). Talking about secondary use, in the late antiquity we are essentially referring to materials from long distance trade, while during the middle Ages local productions were employed, even though case of waste shipped within the same regional area are known (Piombino, Berti & Bianchi 2007) and storage vessels from the Iberian Peninsula were recycled as well (Francovich & Gelichi 1984). Concerning the building technique in itself, it is worth noting that during Late Antiquity the fill was placed in a few levels, while in the Middle Ages pottery was arranged in up to five levels (as in San Domenico in Prato, Vannini 2001). This change is strictly related to the fact that the vessels employed were not made on purpose but recycled; thus, only certain forms could readjust to the shape of the vault at a certain level. The assemblages and the written evidence mentioned in this paper meant to be examples, but the technique was spread all around the western Mediterranean.[9]

---

[9] For a more complete case study on the Iberian Peninsula, especially on the Catalan areas see Caroscio in press.

Acknowledgements

I would like to acknowledge Dragos Gheorghiu and Phil Mason for inviting me to contribute to their session at the 15th EAA Annual Conference. The help of Joseph V. Lerma in getting access to the assemblage of the vessels of the "Convento de Santo Domingo" kept in the "Museo de la Cerámica de Manises" that he directs was priceless. Finally, I would like to acknowledge Jaume Coll, Grazie Berti, Alessandra Molinari, Marco Gentile, Teresa Ribelles and Laura Venturini for bibliographical references and access to the material needed.[10]

Bibliography

AMIGUES, F., CRUSSELLES, E., GONZALEZ-VILLAZSCUSA, R., LERMA J. V.
 1996, Los envases de Paterna/Manises y el comercio bajo medieval, in *La Céramique en Méditerranée Occidentale*, Rabat: Institut Nacional de Sciences de l'Archéologie et du Patrimoine (1991): 346-361.

ARNS, R. G., CRAWFORD, B. E.
 1995, Resonant Cavities in the History of Architectural Acoustics, *Technology and Culture*, Vol. 36, N° 1, Baltimora: Johns Hopkins University Press: 104-135.

BASSEGODA NONELL, J.
 1983, *La cerámica popular de la arquitectura gótica*, Barcelona: Ediciones de Nuevo Arte – THOR.

BERTI, G.
 1990, Pietrasanta. Ceramiche toscane nel recupero di S. Agostino dei secc. XIV-XVII dal Museo Archeologico, in G. Bojani (ed.), *Ceramica toscana dal medioevo al XVIII secolo*, Monte San Savino: 292-322.

BERTI G.
 2005, *Le ceramiche ingobbiate "graffite a stecca". Secc. XV-XVII (Museo Nazionale di San Matteo)*, Ricerche di Archeologia Altomedievale e Medievale, 29, Firenze: All'Insegna del Giglio.

BERTI, G.
 2007, Le ceramiche di Sant'Antimo nel quadro delle importazioni e delle produzioni locali di Pisa del XIII secolo, in G. Berti and G. Bianchi (eds.), *Piombino. La chiesa di Sant'Antimo sopra I canali. Ceramiche e architetture per la lattura archeologica di un abitato medievale e del suo porto*, Firenze: All'Insegna del Giglio: 369-384.

BERTI, G., & BIANCHI, G.
 2007, (eds.), *Piombino. La chiesa di Sant'Antimo sopra I canali. Ceramiche e architetture per la lettura archeologica di un abitato medievale e del suo porto*, Firenze: All'Insegna del Giglio.

BERTI, G., & TONGIORGI, L.
 1974, Coppi del XVI secolo per riempimenti di volte, *Antichità Pisane* I/4: 6-12.

---

[10] martacaroscio@gmail.com
Fellow in Museum Studies
Museo Nacional de Cerámica "González Martí", Valencia (2008-2010)

BLUME, F.
1868, Edictus Langobardorum. Capitula extra edictum vagantia. I. Grimoaldi sive Liutprandi Memoratorium de Mercedibus Commacinorum, *Monumenta Germaniae historica,* Leges IV: 176-180.

BUSELLI, F.
1965, *Palazzo Mediceo a Saravezza,* Empoli: Barbieri-Noccioli.

CAROSCIO, M.
2009, Storage vessels in the Mediterranean area in the late middle Ages. Interpreting technical devices by ethnographic and historical sources, in D. Gheorghiu (ed.) *Experimenting the Past,* BAR in press.

CAROSCIO, M.
in press, Vessels used for shipping goods in the Western Mediterranean during the late Middle Ages and early Modern period, *Medieval Ceramics,* 31.

CARRERAS Y CANDI, F.
1913, Les obres de la catedral de Barcelona (1298-1445), Continuació, *Boletín de la Real Academia de Buenas Letras de Barcelona* XIII, n° 50, Barcelona: Imprenta de la Casa Provincial de Caridad: 128-136.

CARRERAS Y CANDI, F.
1914, Les obres de la catedral de Barcelona (1298-1445), Continuació, *Boletín de la Real Academia de Buenas Letras de Barcelona* XIV, n° 53, Barcelona: Imprenta de la Casa Provincial de Caridad: 302-317.

CUTERI, F. A., & DI FEDE, E.
Bacini e vasi acustici nelle chiese del territorio di Megara (Attica-Grecia), paper presented at the AIECM2 in Venice 2009 (in press).

DIES CUSI, E., & GONZALES VILLAEZSCUSA, R. J.
Las tinajas de transporte bajomedieval y sus marcas de alfarero. Archaeological Report (informe), Servicio de Investigación Arqueologica Municipal de Valencia, unpublished.

FICHERA, G.
2007, Lo scavo del riempimento della volta, in G. Berti and G. Bianchi (eds.), *Piombino. La chiesa di Sant'Antimo sopra I canali. Ceramiche e architetture per la lattura archeologica di un abitato medievale e del suo porto,* Firenze: All'Insegna del Giglio: 149-158.

FRANCOVICH R., & GELICHI, S.
1984, *La ceramica spagnola in Toscana nel bassomedioevo,* Quaderni dell'Insegnamento di Archeologia Medievale della Facoltà di Lettere e Filosofia dell'Università degli Studi di Siena, 3, Firenze: All'Insegna del Giglio.

FRANCOVICH, R., & VALENTI, M.
2002, *C'era una volta. La ceramica medievale nel convento del Carmine,* Catalogo della mostra (Santa Maria della Scala 25 giugno-15 settembre 2002), Firenze: Polistampa.

GONZÁLEZ GONZALO, E.
1987, La cerámica bajomedieval de la Catedral de Mallorca, in *Arqueologia Medieval Española,* II Congreso, Madrid 19-24 January 1987, Vol. III: 470-482.

MOLINARI, A.
2003, La ceramica medievale in Italia ed il suo possibile utilizzo per lo studio della storia economica, *Archeologia Medievale,* XXX: 519-528.

MONNERET DE VILLARD, U.
1920, Note sul memoratorio dei maestri commacini, *Archivio Storico Lombardo* XLVII (I and II): 1-16.

POISSON, J. M.
2005, L'uso dei recipineti ceramici nell'architettura antica e medievale: alcuni esempi in Italia ed altrove, *Archeologia dell'Architettura* X: 55-64.

VANNINI, G.
2001, Una struttura edile trecentesca: il complesso fittile del S. Domenico di Prato, in E. De Minicis (ed.), *I laterizi in età medievale. Dalla produzione al cantiere. Atti del Convegno Nazionale di Studi (Roma, 4-5 giugno 1998),* Roma: Edizioni Kappa: 199-212.

VITRUVIUS, M. P.
2008 (ed.), *De architectura: libri X,* Roma: Edizioni Studio Tesi.

The Reuse of Materials during the Medieval and Post-Medieval Periods: A Case Study of Recycling Building Materials in Rothwell, near Leeds, England

George Nash

Abstract

Research undertaken by the author over the past fifteen years has revealed that many buildings and structures that date to the medieval and post-medieval periods incorporate materials that have been recycled. Recycling usually includes the reuse of brick, stone and timber. Between 2005 and 2007 the author was involved in a large urban regeneration project in the mining town of Rothwell, south of Leeds, Yorkshire. The project included the relocation of a supermarket in Commercial Street to land that was formally occupied by the town's gas works. Prior to relocation of the supermarket and under the auspices of non-statutory Planning Policy Guidance Note No. 15 the author embarked on a desk-based assessment of the town centre as well as a comprehensive historic building survey programme. During this survey programme, which included the detailed study of 12 buildings and a series of ancient boundary walls, it became clear that a systematic programme of recycling materials, in particular medieval roofing timbers had occurred.

In this paper the author discusses the contextual nature of these recycled materials and suggests that the process of reuse during this period was widespread. Observations made within the town demonstrate the need to re-evaluate historic buildings as external elevations may not reveal their true age.

Introduction: The Practice and the Scales of Recycling

Planning a staged approach to investigate the built heritage

The small mining town of Rothwell, south of Leeds is considered an important archaeological and historic resource (Brown 1987; Bulmer and Brown 1999; Green 2001). Over the past ten years the town centre has been the subject of systematic and targeted archaeological investigations that have included intrusive and non-intrusive investigations (Gifford & Partners 2003, 2006; West Yorkshire Archaeology Service 2001, 2002). The focus of several of these investigations has been the relocation of a supermarket which was constructed at the western end of Commercial Street to 'brown-field' land north of Commercial Street that once housed the town gasworks (Gifford & Partners 2003, 2006). This parcel of land contained a number of surviving medieval landscape components, including the rear plots that belong to buildings that front the northern line of Commercial Street (Fig. 1). This irregular parcel of land was, until the recent development, bounded to the west by Church Street, to the south by Commercial Street, to the east by the residential complex at Blackburn Court and to the north by Ingram Parade. The site, prior to demolition comprised a series of derelict 19th and 20th century buildings that occupied the remnants of a series of probable medieval burgage plots, as shown on the Plan of the Township, dated 1839 and the 1st Edition Ordnance Survey of 1894 (Figs 2 and 3).

As part of the planning process Leeds City Council granted planning permission and Conservation Area Consent in September 2006 (Refs: P/06/02494/FU & P/06/02683/CA) to demolish the existing [Morrison] supermarket, neighbouring 20th century shopping arcade, warehouses, factory units and outbuildings (associated with the rear plots that back onto buildings to the rear of Commercial Street) and to erect a new supermarket complex, additional retail units and car parking provision. Prior to demolition, all buildings/ structures within this parcel of land were assessed and recorded using criteria set within English Heritage's *Understanding Historic Buildings: A guide to good recording practice* (2006).

The establishment of the town Gas Works in the north-eastern portion of the development area between 1839 and 1875 would no doubt have resulted in the destruction of potential archaeological remains, in particular those present in the northern sections of the burgage plots that were associated with buildings on Commercial Street. Extensive ground remediation of the former gas works was undertaken by Leeds City Council and, again, this will have compromised the survival of archaeological remains. Later use of the site, until groundworks commenced for the supermarket development, included the creation of a public car park that extended between the Oulton Beck to the north and the former northern boundaries of the properties along Commercial Street.

Within the boundary of the supermarket development were five buildings and associated complexes that were assessed and recorded (Fig. 4) to include:

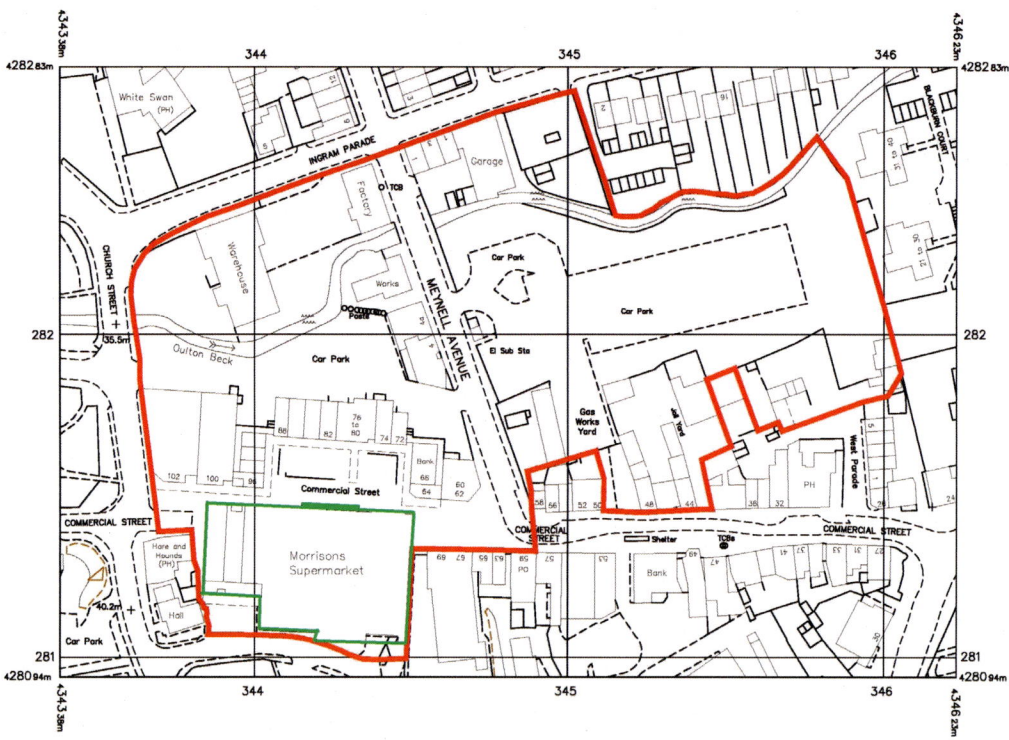

Fig. 1: *The centre of Rothwell showing the location of the former supermarket and the development area (after Gifford & Partners 2003).*

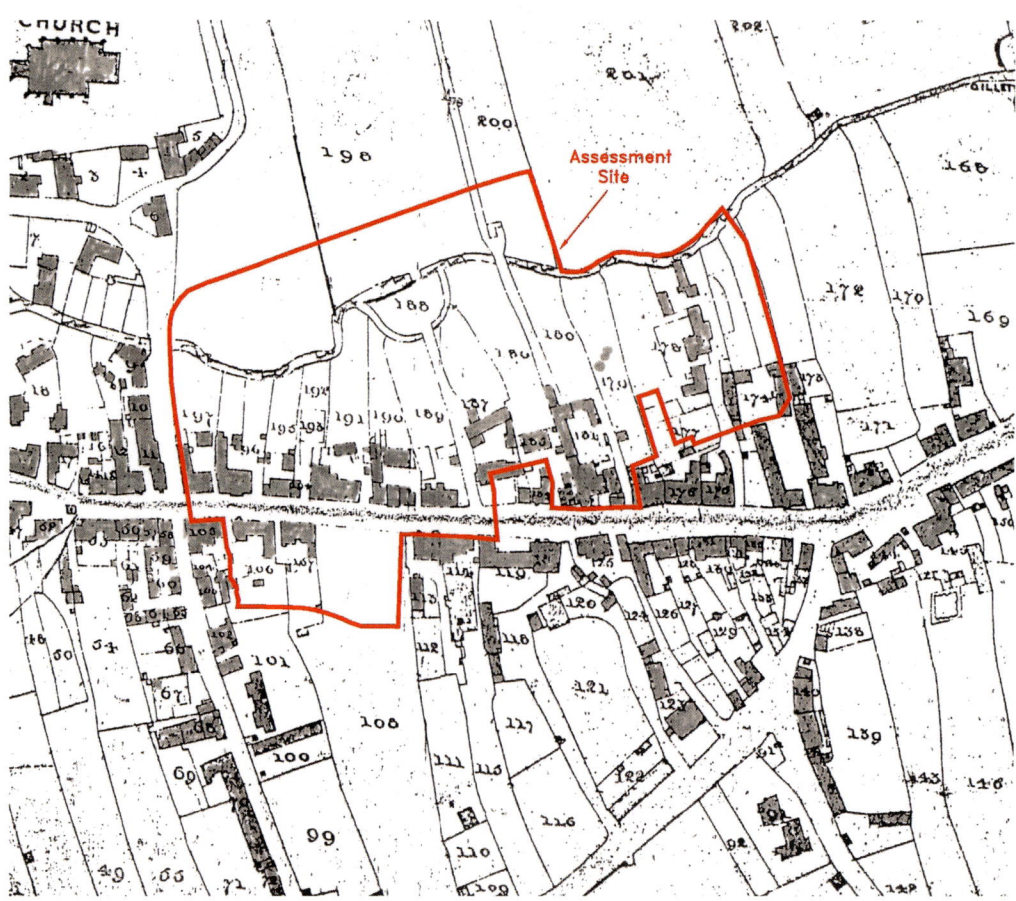

Fig. 2: *Plan of the Township of Rothwell dated to 1839 (the extent of the supermarket development is shown in red).*

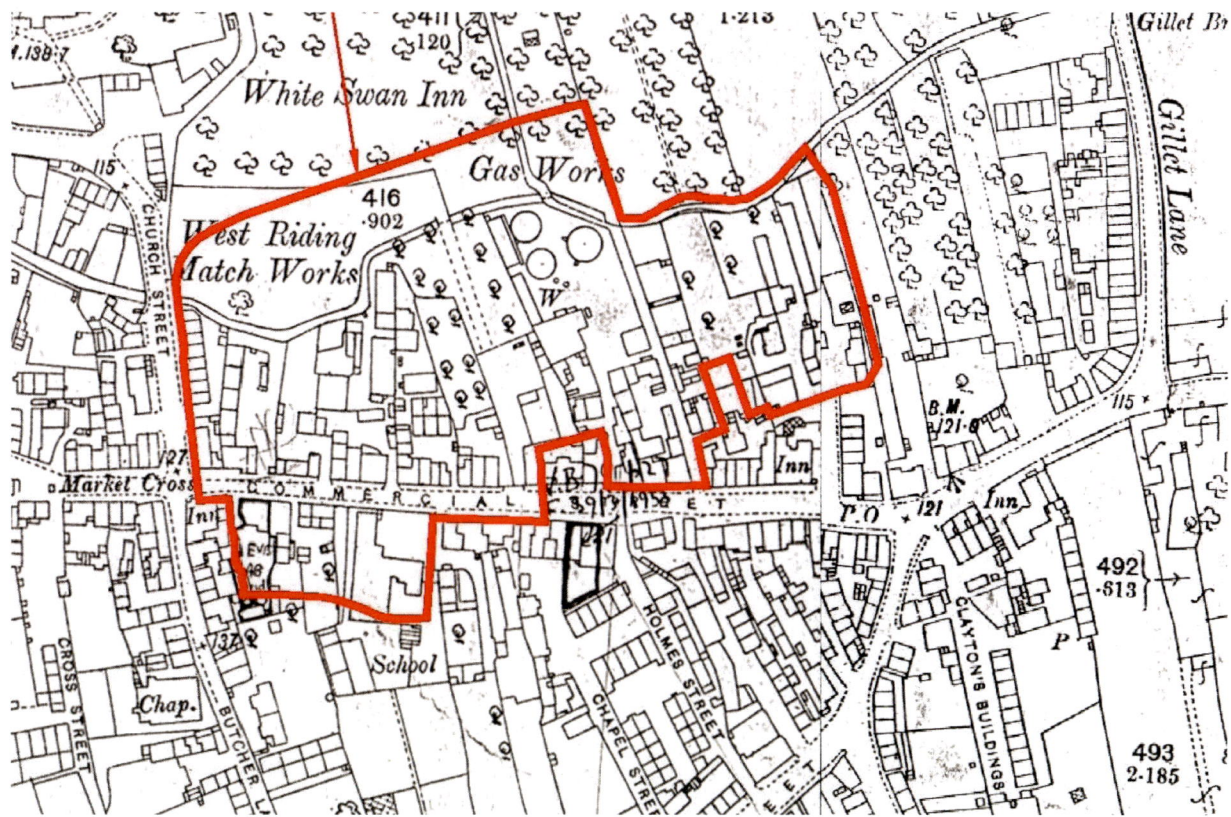

Fig. 3: *The 1st Edition Ordnance Survey sheet of 1894 (the extent of the supermarket development is shown in red).*

Fig. 4: *Map showing the extent of the development and the building complexes and boundary plot walls that formed the assessment (after Gifford & Partners 2003).*

- Two derelict semi-detached residential properties constructed of stone that lie to the north of West Parade at NGR SE 3460 2819;
- A stable block and a series of associated brick-built units;
- A former non-conformist chapel to the east of the Jail Yard to the rear of 32-36 Commercial Street at NGR SE 3455 2819;
- Outbuildings that form Jail Yard at NGR SE 3453 2818; and
- A series of stone outbuildings that lie to the west of Jail Yard, forming a building complex to the rear of No 50-58 Commercial Street.

In addition to these buildings, the author undertook an audit of other buildings and structures, in particular brick and masonry boundary plot walls, lean-to buildings elevations and multi-phased wall sections that had been incorporated into earlier buildings or were included in later building activity. All these contained clear evidence of recycling, using materials that probably originated from Rothwell's medieval past.

As part of the cultural heritage component a staged-approach towards the archaeology was employed, commencing with a desk-based assessment which was published in 2003 (Gifford & Partners 2003). This stage revealed the probability of the survival of medieval remains within the centre of the town.[1] This was later supported by a reconnoitre by the author who managed to gain access to several buildings that, from the outside displayed distinct mid-19th century architectural traits but internally clearly contained medieval components such as floor and roof timbers. Further evidence was revealed with the recording of several wall sections that part-delineate medieval burgage plots to the north of Commercial Street (e.g. at Nos. 56 and 58 Commercial Street). Incorporated into these boundary structures were many dressed medieval stones and early post-medieval brick.[2]

Establishing an historical discourse: Medieval and Post-medieval development

Based on the available documentary evidence, a settlement at Rothwell was probably established in the Anglo-Saxon period, if not earlier, although the history of the area in the 500 years after the collapse of Roman military and administrative control is largely absent. However, it has been reported that an early medieval stone head was found in the graveyard of Holy Trinity Church (SMR/HER ref. PRN 4988[3]).[4] To the east of gasworks, 11th century pottery was recovered from a watching brief, which suggests that the early medieval settlement could have been extensive (WYAS 2001). Despite the lack of evidence for a manorial complex belonging to this period (Green 2001: 7) it is probable that the site of the eleventh century manor known as Rothwell Castle (SMR ref. PRN 1977) may have been built on the site of an early medieval precursor (Gifford & Partners 2003).

The first documentary evidence for Rothwell is recorded in the Domesday Survey of 1086 (Morgan 1983), when during this time the settlement and surrounding landscape was in the ownership of *Ilbert de Lacy*, a prominent Norman figure whose family also held lands in the Welsh Marches and Ireland. Based on the Domesday inventory, Rothwell was at this time a small hamlet, valued at five vills (Appendix 2); the extent of this settlement though is unclear. It is probable though that the former settlement was organised around the parish church (around Holy Trinity Church).

Following the Norman Conquest of 1066 Rothwell's history becomes less obscure with records for a settlement within the manorial grave accounts and messuage entries; a result of the settlement becoming the administrative centre during this time (Green 2001: 4). It has been suggested that during this time the settlement had developed to the east of the manor, roughly where the centre of present-day Rothwell stands (Green 2001: 8; Gifford & Partners 2006); based on the author's reconnoitre of the area this assumption is more than likely. Along with the manor, the site of an 11th century castle, later becoming a fortified manor house - Rothwell Castle lay to the west of the town centre (SMR ref. PRN 1977). It is more than probable that the development of a commercial centre for Rothwell grew either side of what is now known as Commercial Street, a linear east-west thoroughfare that converged with Market Cross, Butcher Lane and Church Street in the west and Gillet Lane to the east. Evidence of this early street is recorded with the presence of 13th and 14th century components incorporated into later buildings that stand either side of this thoroughfare (Plate 1). To the east Commercial Street an archaeological watching brief at Oulton Lane recovered a number of contextually residual pottery sherds dating between the 11th and 18th centuries as well identifying the remains of a post-medieval tannery. Despite the residual nature of the sherds one can suggest that the medieval settlement extends eastwards beyond the current extent of Commercial Street.

---

[1] I will add here that the medieval church of Holy Trinity and the [adjoining] remains of a manor house exist to the north-west of the town centre.

[2] Sometimes referred to as Tudor and Elizabethan brick.

[3] SMR – Sites and Monuments Record; now largely replaced by HER – Historic Environment Record. Both records use PRN's – Primary Record Numbers.

[4] For the sake of this paper, the author will use the neutral term 'Early Medieval' rather than 'Anglo-Saxon'. This term considers the generic archaeology and history of the period rather than basing assumptions on ethnicity.

Plate 1: A medieval timber-framed roof covered by a 19th century tiled roof within No. 43 Commercial Street.

During the 17th, 18th and 19th centuries Rothwell continued to expand and grow in prosperity. During this time a new street plan was established which radiated away from Commercial Street in all directions. Apart from nearby coal mining, the former burgage plots either side of Commercial Street began to be redeveloped, thus reducing the size of some of the plots. This is clearly shown with the expansion of the town gasworks during the mid- and late-19th century when a number of plots belonging to the buildings fronting the north of Commercial Street that once extended to the southern bank of Oulton Beck were dramatically reduced, sometimes to the size of backyards (e.g. Nos. 44, 46 and 48 Commercial Street). A similar scenario occurred to plots that belong to buildings that fronted the southern side of Commercial Street.

The Building Audit

The durability of materials

Based on research and observations made at Rothwell three principal materials survive the archaeological record: brick, stone and timber. I would also add metal to this list although no metal objects were considered archaeologically important and as a result were not recorded during the audit. These materials provide both a structural use as well as an aesthetic value to each of the buildings they were home to. This can be witnessed with the use of shaped and moulded timber panel casements located on buildings such as Nos. 50 and 52 Commercial Street (Plate 4). Besides these principal materials, there is one building complex – Nos. 34-36 Commercial Street that has on its ground floor a symmetrically decorated plaster mounded ceiling (Plate 5). However, neither the plasterwork nor the moulded timber panel casements can be considered forms of recycling, although the construction phases for each building suggest that other components – architectural and structural may have been incorporated during the life of each building.

The following inventory discusses in detail the five building complexes that contained recycled materials (labelled Building(s) 1 to 5 – see Fig. 4). Recovered during the demolition of Building Complex 1 were a number of floor and roof timbers that were clearly medieval in form. However, the precise original location of these timbers is not known.

Building Complex 1: two derelict semi-detached residential properties constructed of stone that lie to the north of West Parade

This building complex comprised two stone dwellings (forming a small terrace) and associated outbuildings. The building stock was incorporated into two adjoining garden plots, located north-west of Commercial Street

The prosperity of the town appears to have extended into the post-medieval period when at that time Rothwell had become a modest-sized settlement. By the 16th century coal was being mined on a small industrialised scale. This discovery in turn provided the power necessary to run small-scale ceramic, glass and metal manufacturing, usually located within the many former burgage plots that radiated either side of Commercial Street (Green 2001). As Rothwell embraced the Industrial Revolution, residential settlement extended southwards beyond the town centre towards Rothwell Marsh; thus another important thoroughfare was established - Marsh Street. Buildings (or part buildings) dating from this period include Nos. 32-36 Commercial Street, probably 17th century in date (Plates 2 and 3).[5]

To the west of Nos. 32-36 Commercial Street is the site of Rothwell Jail. It is argued that fabric of the original jail buildings were incorporated into several buildings to the rear of Commercial Street, within a side street known as Jail Yard (Brown 1997: 26; Gifford & Partners 2006). A substantial quantity of dressed sandstone was also incorporated into several boundary walls that part-delineate the extent of medieval burgage plots to the rear of No. 32 Commercial Street.

---

[5] Designated a Grade II Listed Building.

Plates 2 and 3: *The Commercial Street frontages and rear sections of Nos. 32-36 Commercial Street.*

and West Parade which originally provided access to this terrace. A further access via an un-named un-metalled road led to the rear of both properties. Based on 19th and 20th century cartographic evidence, this terrace had changed little, except for the addition of several lean-to extensions (Plate 6). The stonework on the rear elevation showed signs of different weathering regimes and/or different types of sandstone used; several of these sections probably originating from other buildings (Plate 7). At either ends of the terrace, and flushed into the gable ends, were polychrome brick chimney stacks that were probably later than the original construction of the building. Located within the central part of the roof were two further chimney stacks, each constructed of brick.

Following demolition of this terrace in 2006 a selection of timbers were removed and stacked within the boundary of the site. Inspection of these timbers by the author revealed that the majority were of medieval

Plates 4 and 5: *Moulded timber floral casement panel and plaster moulded ceiling.*

date and could be identified as specific architectural components within the timber-framing tradition and included several ceiling trusses, collars, purlins and several sections belong to an A-frame. The research question arises as to why such a wealth of medieval timber should be incorporated into a 19th century building. Inspection of the external elevations did reveal possible earlier masonry sections but these were in no way medieval in date (the discussion of timber is later assessed in this paper).

Building Complex 2: A stable block and a series of associated brick-built units that lie to the north of No. 30 Commercial Street

Located east of the building terrace and West Parade were a series of outbuildings that belonged to a currently derelict and decommissioned public house (No. 30 Commercial Street). The buildings that immediately adjoin the main building are not affected by the supermarket development, however they are worthy of

Plates 6 and 7: *Front and rear elevations of a 19th century terrace, north of West Parade which housed the medieval timbers.*

comment in terms of the multi-phasing and the reuse of materials. It is clear that these buildings/structures that were originally incorporated into the burgage plot of No. 30 Commercial Street can be considered an organic development that extends the post-medieval period.

Based on cartographic evidence and architectural style, the public house probably dates to the early 19th century. To the rear of this building are two lean-to brick buildings that are probably contemporary with the main building. The smaller and northern most lean-to was probably once utilised as an outside privy. Incorporated into its northern elevation are sections of an earlier wall, recognised by the brick-type. The lower section is constructed of a complex mortared masonry base which may be medieval in date (Plate 8). Overlying this wall section is a small brick section. The brickwork comprises unfrogged 18th century brick and again this section is the

Plates 8 and 9: *Detail of the multi-phased building components and recycled materials used, building located to the rear of No. 30 Commercial Street.*

result of recycling (Plate 9). Incorporated into the upper section of the stonework and the lower brick coursing is a bricked-in window (using later brick).

To the north of the public house and until October 2006 standing within the rear yard of No. 30 Commercial Street was a purpose-built stable block and adjoining outbuildings (Plate 10). This building range, constructed in an L-shaped pattern was constructed of unfrogged red brick elevations that supported sawn timber frame, covered by a slate tile roof. The main stable block, oriented N-S was accessed from the east, via the remains of a small cobblestone courtyard. Prior to demolition much of the internal features had been removed. However, based on the brickwork on the western elevation, the main stable block comprised three bays. The roof was supported by a series of machine-sawn pitched pine roof timbers. The upper

Plates 10 and 11: *The stable block with recycled cast-iron column and adjoining outbuildings.*

brickwork of the eastern elevation was supported by a single reinforced steel joist (RSJ). Brick piers that protruded from the north and south gables supported this joist. Further support to the eastern section of the stable block was given by a single recycled cast iron column (see Plate 10). It is probable that this substantial piece of architecture may have originated from a nearby industrial building.

Prior to demolition, a smaller range of outbuildings associated with the stable block (and forming an L-shaped annex) existed to the north-east. This series of outbuildings, constructed of red frogged brick included a piggery (Plate 11).

Building Complex 3: A former non-conformist chapel to the east of the Jail Yard to the rear of 32-36 Commercial Street

Located to the west of the stable block and to the rear of Numbers 32-36 Commercial Street was a 19th century non-conformist chapel building (Plate 12). This simple single-phased rectangular building, constructed of

Plates 12 and 13: *The eastern [front] elevation of the former chapel and the recycled northern foundation wall that it once sat upon.*

frogged red brick and supporting a slate tiled roof was demolished in October 2006. The building previously utilised as a community hall and meeting place for the *Royal Antediluvian Order of Buffaloes* (RAOB) was located within a rectangular plot and was approached by a gated boundary to the east. Located within the northeastern corner of the chapel plot is a small square brick outbuilding of unknown use.

The chapel building was not internally inspected and the survival of fixtures and fittings were unrecorded, although two timber A-frames each made of pitched pine were recorded following demolition (Gifford & Partners 2006). The frontage and main approach to the chapel was from the east, via West Parade. The eastern gable comprised a symmetrically laid door and window arrangement. A central door, constructed of pine slating was approached via a narrow stone pathway. The main entrance and windows were recessed and set within a sandstone block surround. The upper section of both the doorway and the windows were architecturally similar, constructed of a stone sill and pyramidal lintel.

A centrally-placed keystone was inserted in each lintel casement. The window openings were encased within brick returns. A central squared stone mullion divided each of the window openings into two equal sections. The style of the window frames could not be ascertained due to all windows being blocked-in.

Following demolition of this building, a sandstone mortared wall extending along the length of the northern elevation was exposed (Plate 13). This wall stood around 0.90m in height and may predate the construction of the chapel and represent the northern extent of a post-medieval or early 19th century plot. The western end of this wall converged with further stone and brick walling that probably also represented late post-medieval plot development.

Building Complex 4: Outbuildings that form Jail Yard

Located to the west of the chapel (see Plate 12) were the buildings that enclose Jail Yard. The buildings within Jail Yard, formally used as the town's gaol, comprised four units; all prior to demolition in October 2006 were derelict (Plate 13). Jail Yard was originally accessed by a narrow alley (or *shut*) leading from Commercial Street (east of No. 44 Commercial Street). The yard, forming a rectangular plot possibly represented two or maybe three medieval burgage plots that would have, prior to the early to mid 19th century extended to the southern bank of Oulton Beck.

Plate 14: *The former jail building located within the eastern part of Jail Yard and the approach to Jail Yard via Commercial Street (looking north).*

Two of the buildings constructed along the western side of the alley that approached Jail Yard remain standing and have recently undergone sympathetic refurbishment. These buildings, now providing rented accommodation were originally outbuildings associated with a butcher's shop that stands in Commercial Street (No. 44). It is probable that both outbuildings, constructed of 19th century unfrogged brick were abattoir buildings (Plate 14). Buildings once occupying the eastern side of Jail Yard probably included the town jail. This two-storey building, constructed of red unfrogged brick and supporting a slate tiled roof appears to have been much altered and, according to 19th century cartographic evidence extended eastwards forming, in plan, a rectangular building. In recent times this building and others within the eastern side of the yard had been cement rendered and therefore their original form was difficult to assess.

The northern boundary of Jail Yard, comprising a mortared stone wall had, in recent times been demolished and rebuilt (e.g. Plate 20). Much of the stone from this wall appears to have originated from buildings within the vicinity. Much of the eastern and western boundary sections that delineated the yard area probably date to the early post-medieval period (based on early 19th century cartographic evidence and building materials).

The northern boundary, as with other northern boundary sections, formed rear plots belonging to buildings that front Commercial Street and were probably 19th century in date and contemporary with the construction of the town gas works that once occupied the car park space to the rear of this and other plots.

The boundary wall that delineated the western extent of the Jail Yard complex provided some insight to the complexity of the plot arrangement within this part of the town. The western boundary oriented N-S and forming the eastern line of an access to the car park, ran to the rear of No. 48 Commercial Street. The wall, constructed in three phases also comprised a substantial red brick section, suggesting at least three phases of construction (Plate 15). This section was located immediately north of No 48 Commercial Street and formed the western plot boundary. The masonry section, the earliest within this boundary was sandwiched by a recently constructed breeze-block wall, and adding to the complex history of the plot development. The red brick section, also forming the western wall of a lean-to building within Jail Yard was 19th century in date. It is probable that this lower section was the original boundary wall that would have (according to the 1st Edition Ordnance Survey map) extended to the Oulton Beck, forming the western

Plate 15: *The three-phased wall section that formed the western boundary wall or the rear plot.*

boundary of a medieval burgage plot. The E-W breeze-block wall section formed the northern boundary and converged with a recently re-laid masonry wall (see Plate 20).

Building Complex 5: A series of stone outbuildings that lie west of Jail Yard, forming a building complex to the rear of No 50-58 Commercial Street

To the rear of Nos. 50-58 Commercial Street were surviving two buildings and a sandstone masonry boundary wall. Based on the Ordnance Survey maps of 1932 there existed a complex arrangement of outbuildings and workshops that formed around a rectangular courtyard. Access to this courtyard was via Commercial Street, between Nos. 48 and 50 Commercial Street. During the mid to late 20th century much of the arrangement of this rear plot complex was radically altered. Surviving until October 2006 were several free-standing walls, once forming part of a lean-to complex of outbuildings and a rectangular workshop, located between Nos. 54 and 56 Commercial Street. During medieval times this rear section of the Commercial Street buildings was divided into at least three burgage plots. In recent times No. 58 Commercial Street had been utilised as a builder's supply yard (pertinent to this paper, scattered around the rear section of the plot were roofing tiles and ornamental brick types).

A sandstone masonry wall extended long the building line between numbers 52 and 54 Commercial Street. To the east of this wall, opposite the western boundary wall of No. 48 Commercial Street and once forming the rear eastern plot boundary to No. 50 Commercial Street was a further multi-phased masonry wall (Plate 16). Inspection of this structure showed clear chiselled dressed medieval stonework. The date of the wall is unclear but was probably late medieval/early post-medieval.

The boundary wall between Nos. 54 and 56 Commercial Street, standing around 2.9m above the ground level of the car park was also once the eastern structural wall for several interconnecting lean-to outbuildings/workshops (Plate 17). Similar to other boundary walls that formed a number of the rear plot boundaries this wall was multi-phased, with several sections dating to the medieval period (based on the diagonal chiselling and pink to yellow lime mortar bonding). The line of this and the boundary wall delineating the western wall of the Jail Yard complex probably represented the surviving southern inner sections of burgage plots that belonged to properties that fronted Commercial Street.

The masonry within the eastern-facing wall section showed evidence of repair and appears to have been constructed in several phases, each representing the extent of former lean-to outbuildings within the plot belonging to Nos. 54/56 Commercial Street. The gabled building, located within the north-eastern corner of this plot was constructed of stone and supported a flagstone tiled roof (Plate 17). The building was accessed via the plot on its western gable. There was also access between this building and the neighbouring outbuilding to the south. There was evidence in both buildings for the reuse of stone. Clearly visible within

Plate 16: *Multi-phased sandstone wall forming the eastern rear wall of No. 50 Commercial Street.*

Plate 17: *Outbuilding/workshop with tiled roof, located to the rear of No. 50 Commercial Street.*

these buildings and the surviving masonry to the south were dressed and worked (chiselled) stone which was probably medieval or early post-medieval in date. The stonework used within the building located within the NE corner of the plot appears to be represented by two sandstone types (Plate 16). The upper section has (through erosion processes) degenerated considerably, whilst the lower has not. The rates of degeneration may be the result of the quality of the stone's geology or one section has been exposed longer than the other. There was evidence of re-pointing in those areas of masonry where erosion has not occurred.

To the west of the north-eastern building (Plate 17) and forming the northern boundary was a 19th century masonry wall, constructed of worked sandstone blocks.

This mortared wall post-dated the outbuilding, standing around 1.5 m in height and was capped by a series of semi-circular [sandstone] copping stones. Some of these had been replaced with more decorative types. The sandstone coursing was regular and continued to form the boundary wall that extended along the western side of the plot (forming the N-S boundary wall of No. 58 Commercial Street and fronting a 20th century tramway thoroughfare known as Meynell Avenue [first shown on the OS map of 1908]).[6]

Defining New Boundaries with Old Materials: The Rear Plot Walling

The northern plot wall

As part of the building audit, Gifford and Partners were required to record in detail a masonry wall that once formed the northern boundary wall to a series of plots that formally belonged to buildings that fronted Nos. 38, 40 and 42 Commercial Street and stood on average 1.15m in height and extended *c.* 20m. Based on the *in situ* masonry located at the base of the wall this structure appears to considerably predate the stable complex (Building Complex 2) and the chapel (Building 3), possibly representing a short section of a former medieval burgage plot boundary wall (Plate 18). The east-west section of the wall is however, probably contemporary with the expansion of the town gas works during the mid to late 19th century, possessing no clear formal *in situ* stone layout. Based on the late 19th century Ordnance Survey mapping this wall probably extended to the southern bank of Oulton Beck (similar to other N-S former plot walls that survived prior to demolition in October 2006). Incorporated into this wall section were a number of masonry types that indicate that stone was robbed from once nearby standing buildings. Masonry types included dressed and worked stone, some with irregular chisel marks. Other stone showed signs of weathering and clearly originated from other buildings.

This predominantly sandstone wall section was demolished in November 2006 following a detailed recording programme that included drawn and photographic records. Also included within this part of the survey was a section of re-built sandstone masonry walling that formed the rear northern boundary line of Nos. 42-44 Commercial Street. This wall section, along with other masonry sections within a rear garden area appears to be recent, probably constructed in the 1980s using medieval stone from nearby demolished buildings (Plate 19). Incorporated into the fabric of the wall were a number of cut stone blocks that had chisel marks scored across the surface. These chiselled blocks arguably date to the medieval period (Brunskill 1978).

Extending westwards from the burgage plot boundary wall and united by a *return* was a later wall section (Fig. 5). Incorporated into this short wall section were

Plate 18: *Eastern section of a rebuilt N-S medieval burgage plot boundary wall containing dressed medieval stone, looking SW.*

---

[6] It is probable that two further N-S walls existed within the area of the present plot, delineating the plots belong to numbers 54 and 56 Commercial Street.

Working with the Past: Towards an Archaeology of Recycling

Plate 19: *Reuse of medieval stone blocking in c. 1980, bonded with Portland cement.*

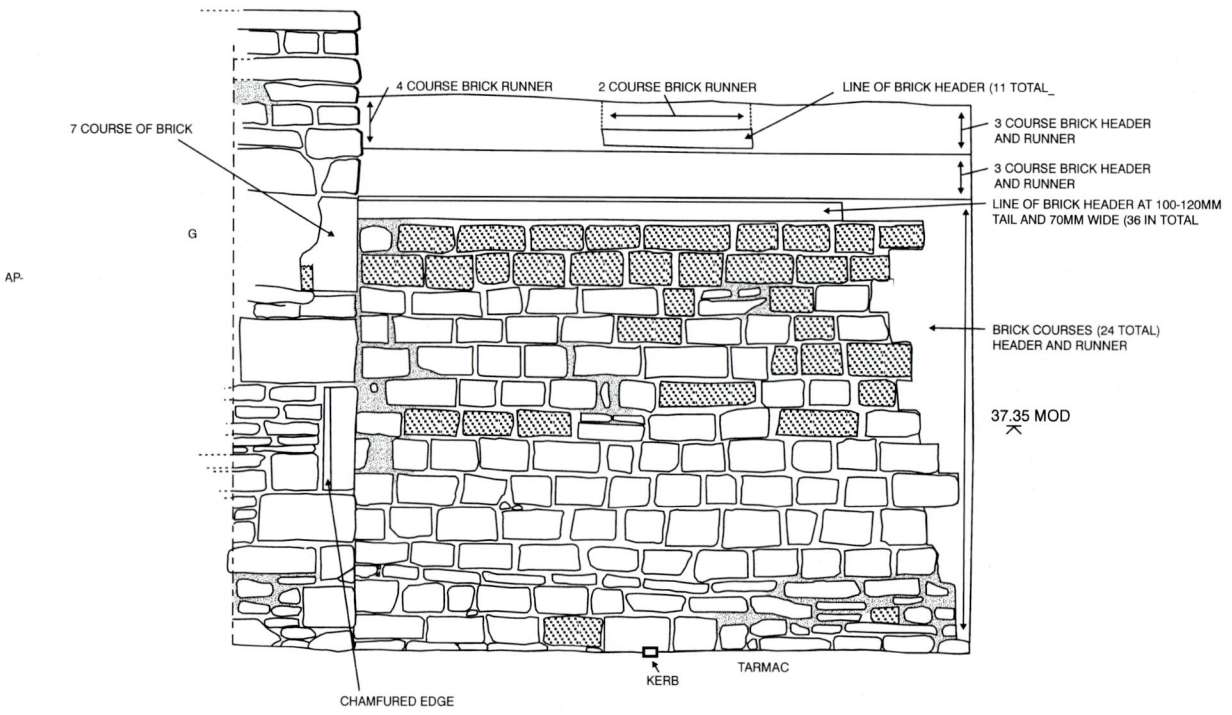

Fig. 5: *A mid 19th century east-west wall section constructed using a variety of materials.*

several brick types, both frogged and unfrogged 19th century brick types which indicate localised repair. This wall section supported a large masonry coping stone that appears to have originated from a more formal garden/boundary wall. The brick wall, constructed of unfrogged brick extended some 4m to the west where it is abutted by a recently constructed [reused] masonry wall (see Plate 19).

Masonry within plot belonging to No. 54 Commercial Street

Following the demolition of the buildings that were located within Building Complex 5 and exposed within the plot belonging to No. 54 Commercial Street were two surviving sections of masonry that originally formed the supporting wall for several outbuildings

Plate 20: *Multi-phased wall section, located between Nos. 56 & 58 Commercial Street.*

(Plate 20). Both surviving sections, constructed of several types of sandstone and standing around 3.3m in height, were in a good state of preservation. Incorporated into the west-facing section of walling and abutting the rear [brickwork] elevation of No. 54 Commercial Street were a number of medieval worked sandstone blocks that had irregular chisel marks. It is clear that these stones may originate from earlier structures. Furthermore, it is probable that this surviving wall, along with the neighbouring masonry wall (to the north) dates to the 17th or 18th centuries. Both walls, as suggested earlier may have delineated an earlier [medieval] burgage plot boundary. The masonry section, immediately north of the wall was constructed similarly. Between both sections is a return constructed of short and long quoins. It is probable, and based on cartographic evidence, that the masonry section that abuts the north-facing elevation of No. 54 Commercial Street was a plot wall, whilst the northern section with its return formed the eastern wall of a [recently demolished] lean-to building. Incorporated into this wall was a small oven or furnace and internal chimney stack, both were constructed of red frogged brick (Plate 21). This structure clearly post-dates both walls, probably early- to mid-20th century in date. Beyond this structure, to the north, the masonry wall had been demolished.

Plate 21: *Small oven attached to the boundary wall section between Nos. 56 and 58 Commercial Street.*

Working with the Past: Towards an Archaeology of Recycling

Old for new: The timber assessment

Given the value of wood as a building material, the reuse of timber framing appears to have been widespread (Brunskill 1978; Harris 1993). Recovered from the roof of Building 1 (the terrace located to the north of West Parade) were 12 hand sawn and adze-shaped timbers (e.g. Plates 22 and 23). Accompanying this group of timbers was a number of machine-sawn timbers. The machine-sawn timbers probably date to the late 19th or early 20th centuries, whilst the 12 hand-cut timbers most probably date to the medieval period (c. 14th/15th century). As a result of this discovery all medieval timbers were recorded with particular detail to timber shape and joint type (Figs 6 to 9). From this assessment all timbers were classified into structural type and function.

Plate 22: *Medieval timbers stacked in a corner of the development area.*

Plate 23: *Mortice and tenon joint chiselled out of a roof collar.*

Initial inspection of the timbers suggested that the majority originate from a number of buildings (rather than a single building). It is clear that medieval structures exist within Rothwell and the timbers from this terrace are clear evidence of this. It was considered that several timbers (Timbers 3 and 4) were worthy of dendrochronological analysis. These timbers were inspected and samples [slices] were taken. The samples later proved inadequate due to the absence of sapwood.

All timbers were of oak and were identified, including three floor joists with complex mortice and tenon joints on Timbers 2, 4 and 8. Many joints are accompanied by (bit and brace drilled) peg holes. Timber 3 probably represents the left section of a floor joist. All joists are substantial in size and weight. Timber 4 was by far the most substantial and complex in terms of jointing. This roof timber, extending in length to 5.2m would have been regarded as key roof timber, probably supporting a number of vertical timbers (e.g. the queen and king posts), whilst employed as a major floor timber.

Timber 2, measuring 4.2m, was chamfered on two of its edges, was made of oak and possessed no jointing. Probably originating from a later building, this timber was probably used as an extended collar or tie beam, supporting a series of floor joists.

Carpenters marks (also referred to as assembly marks) were present on six timbers (Timbers 5, 6, 7, 8, 9 and 10). The marks were usually in form of Roman numerals (e.g. Timber No. 9; Plate 24). There were several marks on Timbers 6 and 7 that were not Roman numerals but assembly marks that indicate major joints. The location of all carpenters marks suggests the likelihood that each timber formed part of an assembled (prefabricated) frame.

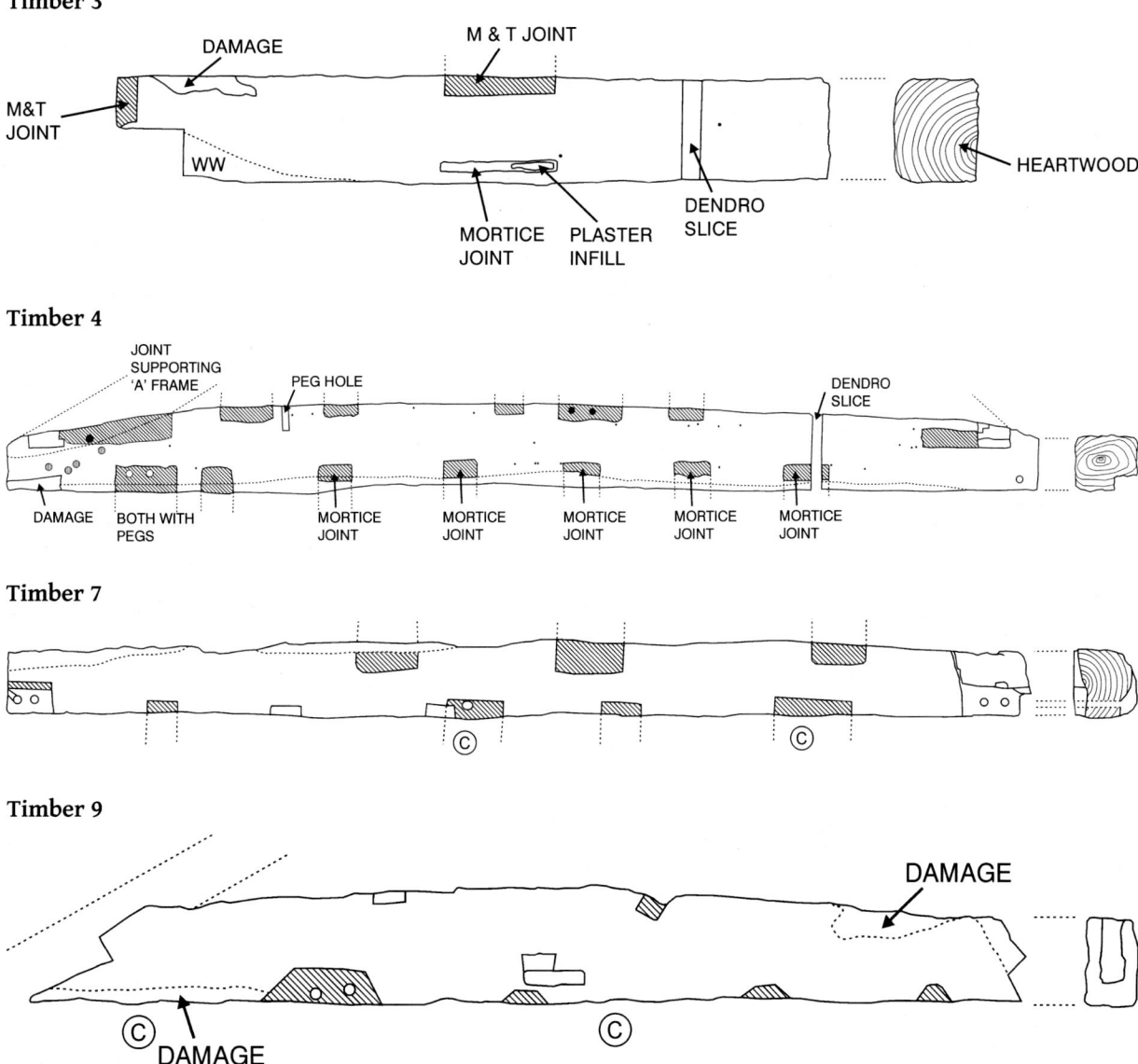

Figures 6 to 9: *A selection of recorded medieval timbers (ceiling joints) from a terrace north of West Parade; (© - carpenter's mark).*

Plate 24: *Carpenters/assembly marks on Timber 9.*

Recycling brick and stone

A brick and stone audit was undertaken in November 2006. Where possible traceable brick and stonework was inspected and recorded photographically. However, it should be noted that the demolition/groundworks team removed off site much of the building debris including large quantities of stone and brick (this material has a considerable recyclable value). The material that remained was dragged into a heap within an area that once formed Jail Yard. The origin and location of the material though is not known. However, identifiable medieval and post-medieval dressed and worked stone was recorded from the skip (Plates 25 and 26). Also included within this audit were a number of copping stone types and chiselled worked stone which probably originated from local quarries.

Plate 25: *Medieval dressed sandstone with chisel marks on the external face.*

Much of the stonework remaining within the bounds of the development site comprised three types: dressed, worked and roughly hewn material. Fortunately, the masonry that formed the northern boundary wall and the N-S oriented wall belonging to the rear plot of No. 54 Commercial Street gave some insight to its possible function and it is clear, based on the reconnoitres undertaken by the author between 2004 and 2007 that much of the masonry was reused and either incorporated into the fabric of the outbuildings or within the boundary walls that delineated the rear plots of Commercial Street.

Plate 26: *Post-medieval dressed stone, possible copping stone or mounded casement.*

The brick audit identified eight brick types, the earliest dated from the 16th/17th centuries comprising a hand-made unfrogged part vitrified type (Plate 27). A similar brick type was recorded within a small outbuilding unit located to the rear of No. 30 Commercial Street. Other types included several unfrogged types which probably date to the early to mid 19th century and probably originate from the stable block (Building 2) or from one of the brick buildings within Jail Yard (Building Complex 3).

Frogged brick types were by far the most numerous found on site. Many of these types are stamped with maker's marks such as Cliff & Sons [Leeds] (Plate 28). These brick types date from the mid-19th century and were incorporated into a number of buildings that are within the bounds of the site.

Plate 27: *Probable 16th/17th century part vitrified brick.*

Plate 28: *Corner brick from the Cliff & Son factory in Leeds, late 19th century in date.*

Discussion: Reuse, Recycle and Renovate

Despite the sometimes ad hoc reuse of materials during the medieval and post-medieval periods there appears to be an acceptance to observe and respect former burgage plot boundaries and the footprints of former buildings. The medieval builders and plot owners of Rothwell, not unlike those from other medieval urban centres elsewhere in Britain and Europe had, due to the fluctuating economics of the day, needed to utilise existing materials either from buildings that were in use (i.e. rebuilding) or to obtain materials from buildings that were either derelict or abandoned. During the post-medieval period and the 18th and 19th centuries, the demands and shortages for core building materials, in particular stone and timber forced builders to make-do with the material available.

The present study on the buildings that were demolished to make way for the new supermarket, show that rebuilding and refurbishment were essential devices in upgrading and altering early building regimes. An important and common-sense approach to these sometimes radical building projects was to carefully select recycled materials. As I have demonstrated, recycled brick, stone and timber formed an essential resource for any building project. Due to the hidden nature of these rebuilding projects (i.e. to buildings and structures to the rear of Commercial Street) builders were not too concerned with the visual aesthetic value of each building project. The same cannot be said for the Commercial Street frontages where the ethics of rendering, facade symmetry and cleanly laid brickwork were the norm.

The formal building plan of one architectural regime, say, that of the medieval timber framing tradition appears to be an acceptable addition to a later and completely different building type such as the 19th century stone and brick terrace. Although, the later building type is arguably more formalised and uniform in its building methodology, the builders of the day considered the recycling and incorporation of a series of medieval timbers into their building as an acceptable construction practice.

The (re)building projects employing the recycling of materials span the medieval and post-medieval periods and I would argue that Rothwell is not the only urban centre to employ such a regime. Throughout both periods, the historical record informs us that those urban centres such as Rothwell experience periods of both economic decline and prosperity. It is more than probable that during times of decline recycling materials was an essential process of economy.

Interestingly, and reflecting the current trends on design and build, no direct recycling programme was employed between Rothwell's former Commercial Street supermarket and the new one. However, despite this the materials from the buildings that once occupied the land north of Commercial Street along with brick and concrete from the former supermarket were crushed, graded providing a secure foundation mat for the new supermarket which opened its doors to the public in 2007.

Acknowledgements

I would like to thank the editors for inviting me to publish this material and also to Caroline Malim for producing such superb finished drawings of the timbers. I would also like to thank DLA Architecture (Bradford) and Paul Kettlewell of Morrison Supermarkets PLC for allowing me to publish this paper. Finally, I would like to thank Abby George for her useful comments. All mistakes are of course my responsibility.

Bibliography

BROWN, A.
    1987 *Rothwell in the 900 Years after Domesday.* Rothwell Advertiser Press Ltd.
    1997 *Albert Brown's Story of Rothwell.* Stephen Ward Photography and Publishing, Rothwell.

BRUNSKILL, R. W.
    1978 Distribution of building materials and some pln types in the domestic architecture of England and Wales. *Transactions of the Ancient Monuments Society of England and Wales (New Series)*: 41-66.

BULMER, S. and BROWN, A.
    1999 *Images of England: Around Rothwell.* Stroud: Tempus Publishing Ltd.

ENGLISH HERITAGE
    2006 *Understanding Historic Buildings: A guide to good recording practice.*

GIFFORD & PARTNERS
    2003 *Commercial Street, Rothwell, West Yorkshire: Desk-based Assessment.* Report No. 10687, R01.
    2006 *Commercial Street, Rothwell, West Yorkshire: Building Survey to the rear of Commercial Street,* Report No. 10687, R05.

GREEN, D.
    2001 *Rothwell - Historic Settlement Assessment Report.* Unpublished Report prepared for West Yorkshire Archaeology Service.

HARRIS, R.
    1993 *Discovering Timber-framed Buildings.* Oxford: Shire Publications.

MORGAN, P. (ed.)
    1983 *Domesday Book: Yorkshire.* Chichester: Phillimore & Co.

SMITH, A. H.
    1961 *The Place-Names of the West Riding of Yorkshire. Part II.* Cambridge: Cambridge University Press.

WEST YORKSHIRE ARCHAEOLOGY SERVICE (WYAS)
    Nd. *Field Names: West Yorkshire Tithe Awards (M - R).* Unpublished Report.
    2001 *Hannah House, Oulton Lane, Rothwell, West Yorkshire: Archaeological Watching Brief.* Unpublished Report, No. 917.
    2002 *Land Adjacent to Rothwell Manor, Rothwell, West Yorkshire: Geophysical Survey.* Unpublished Report. No. 1042.

APPENDIX 1

Cartographic sources consulted

1755. Thomas Jeffry's Map

1783. Copy of the Plan of the Township of Rothwell Haigh in the Parish of Rothwell made to Correspond with the award dated March 1783. John Sharp & Richard Clark, Commissioners.

Allerton Park 15. WYAS(W)

1786. A Plan of the Township of Rothwell in the County of York including Royds and the whole of Rothwell Haigh made from an Actual Survey taken in 1786 by order of Charles Brandling Esq. The owner of the impropriate Rectory of Rothwell

1816. A Plan of the Township of Rothwell with Royds and Outon with Woodlesford in the parish of Rothwell in the County of York made in 1816 by order of John Blayds Esq. By Henry Teal of Leeds. Farrer 06/b WYAS(W)

1819. Enclosure Plan for Rothwell. Commissioners. Henry Teal and Richard Clark

1819. Enclosure Plan of Stone Bridge Green and Rothwell Marsh, Henry Teal, Surveyor; A37 WYAS(W)

1834. Plan of Part of the Estate belonging to John Blayds Esq. In the Parishes of Methley, Rothwell and Wakefield, DB/M/208, WYAS (L)

1839. Plan of the Township of Rothwell, DB/M153 WYAS(L)

1839. Plan of an Estate Situate in the Township of Rothwell the Property of Mr John Hindle and Others, DB/M232 WYAS(L)

1842. Ordnance Survey Map, Sheet 233 NE, Scale 6" to 1 mile

1844. Plan of the District of Rothwell-cum-Royds in the Parish of Rothwell and the County of York (Tithe Map by Richard Gouthwaite) RD/RT/201, WYAS(L)

1875. The Village of Rothwell Near Leeds being Part of the District of the Rothwell Local Board ACC 2304/6 WYAS(L)

1894. Ordnance Survey Map, Sheet 233 NE, Scale 6" to 1 mile

1908. Ordnance Survey Map, Sheets 233.3 and 233.4, Scale 25" to 1 mile

1908. Ordnance Survey Map, Sheet 233 NE, Scale 6" to 1 mile

1921. Ordnance Survey Map, Sheets 233.3 and 233.4, Scale 25" to 1 mile

1931. Ordnance Survey Map, Sheets 233.3 and 233.4, Scale 25" to 1 mile

1932. Ordnance Survey Map, Sheet 233 NE, Scale 6" to 1 mile

1938. Ordnance Survey Map, Sheet 233 NE, Scale 6" to 1 mile

1949. Ordnance Survey Map, Sheet 233 NE, Scale 6" to 1 mile

1951. Wakefield (Leeds): Sheet 78. Solid Geology, Scale 1:50,000

1962. Wakefield (Leeds): Sheet 78. Drift Geology, Scale 1:50,000

1978. Wakefield: England & Wales Sheet 78. Solid and Drift Geology, Scale 1:50,000

APPENDIX 2

Derivation of the place name evidence of Rothwell

The derivation of the place-name of Rothwell probably has an association with a pre-Norman settlement. The Domesday Survey refers to the name of *Rodouuelle* and place-name evidence by Smith (1961: 143) suggests that this early form of Rothwell originates from the Old English *Roð(a)-wella*. The elements of the word most likely originate from the Old English *roð*, meaning *clearing* and *wella* representing *well*, thus meaning the well by the clearing. These elements probably show that an early settlement existed within a wood clearing, probably of [ethnically] Anglican origin.

APPENDIX 3

The Domesday Inventory of 1086

In Rothwell, Lofthouse, Carlton, Thorpe (on the Hill) and Middleton, 24 carucates and 1 bovate of land taxable, 12 ploughs possible there. Harold, 14c, Barthr, 7.5c: Alric, 10.5 bovates; and Steinulfr, 10.5 bovates, had halls there. Now Ilbert has there 2 ploughs; and 16 villagers and 1 small holder and 8 ploughs. 1 mill, 2s; meadow, 9 acres, woodland pasture, 2 leagues long and 1 wide. He whole manors, 2 leagues long and 2 wide. Value before 1066 £8; now 65s (£0.65p).